"A nuanced look at how we die . . . Cohen's compelling and eye-opening new book explores complex end-of-life issues with clarity and, despite his own strong opinions, an openness to disparate points of view. . . . This debate is tougher even than the impasse over abortion rights."
　　　　　　　　　　　　　　　　　　　　　—*Huffington Post*

"Lewis Cohen tells a difficult, gripping story, which raises end-of-life issues that affect all of us in one way or another. *No Good Deed* is lucid, fair-minded, and compassionate. It is a work of eloquent necessity."　　　　　—Edward Hirsch, author of *Wild Gratitude* and winner of the National Book Critics Circle Award

"Now that medicine is capable of near-miracles in keeping us alive, we enter the quagmire of who decides when enough treatment is enough, how life is stopped, and what constitutes care versus murder. This superb page-turner is a front-row seat to the life-or-death drama unfolding for doctors, nurses, and families nationwide. It's a must-read by a brilliant doctor who straddles the front lines."　　　　　—Marilyn Webb, author of *The Good Death: The New American Search at the End of Life*

"For families with older and ill members, and equally for health professionals, Dr. Cohen's compelling narrative brings to light the ever more complex world of palliative and end-of-life care."
　　　　　　　　—Dennis McCullough, MD, associate professor, Dartmouth Medical School, and author of *My Mother, Your Mother*

"Serious illness is a complicated issue, but the one thing most people can agree on is that they don't want their loved ones to suffer. Lewis Cohen explains the risks and pitfalls doctors and nurses experience as they try alleviate their patients' pain. This book is a beautifully written and compelling page turner."

—Diane Meier, MD, winner of the MacArthur
Foundation Award and director of
the Center to Advance Palliative Care

"*No Good Deed* shines a much-needed light on the cultural chasm that divides Americans on subjects of ethics, dying, and intensive care. Lewis Cohen has issued a wake-up call to society to openly discuss how we care for people at the end of life. We avoid this call at our peril."

—Ira Byock, MD, professor, Dartmouth Medical School,
and author of *Dying Well* and *The Four Things That Matter Most*

Ellen Augarten

About the Author

LEWIS M. COHEN, MD, is an internationally known professor of psychiatry from Tufts University School of Medicine and a clinician-researcher based at Baystate Medical Center in Springfield, Massachusetts. Dr. Cohen was awarded a Guggenheim Fellowship and a Rockefeller Bellagio Residency to complete *No Good Deed*. He has two sons and lives in Northampton, Massachusetts, with his wife.

Lewis M. Cohen, MD

No Good Deed

A Story of Medicine,
Murder Accusations, and the
Debate over How We Die

HARPER

NEW YORK · LONDON · TORONTO · SYDNEY

Though conversations come from my keen recollection of them, they are not written to represent word-for-word documentation; rather, I've retold them in a way that evokes the real feeling and meaning of what was said, in keeping with the true essence of the mood and spirit of the event.

HARPER

A hardcover edition of this book was published in 2010 by Harper, an imprint of HarperCollins Publishers.

HarperCollins books may be purchased for educational, business, or sales promotional use. For information please write: Special Markets Department, HarperCollins Publishers, 10 East 53rd Street, New York, NY 10022.

FIRST HARPER PAPERBACK PUBLISHED 2011.

Designed by Eric Butler

Library of Congress Cataloging-in-Publication Data has been applied for.

ISBN 978-0-06-172177-9

11 12 13 14 15 ID/BVG 10 9 8 7 6 5 4 3 2 1

This book is dedicated to my loving and insightful wife, Dr. Joan Berzoff, and our two stalwart sons, Zeke and Jake

No good deed goes unpunished.

—Clare Boothe Luce

Contents

Author's Note

I would like to emphasize from the outset the focal point of this book is accusations over clinical procedures and decisions considered by most authorities and medical organizations to be both legal and ethical. The book is not about nurses and doctors such as Jack Kevorkian who espouse the practices of euthanasia (which is illegal throughout the United States) or even physician-assisted suicide (which is legal at this moment only in Oregon and Washington—and in Montana if a lower court decision is upheld). It is not about catastrophes such as occurred in New Orleans when Hurricane Katrina devastated Memorial Hospital and left medical staff isolated without basic resources to care for suffering and terminally ill patients. Neither is it about psychopathic medical professionals who are bona fide serial killers. Each of these separate issues is germane and will be touched upon, but they are not central.

My primary concern is directed at the well-meaning clinicians who believe they offer reasonable end-of-life care and are then shocked to find themselves accused of murdering or euthanizing patients. I am writing about health care professionals who thoughtfully withdraw or withhold life support treatments and provide analgesic medications to alleviate pain and suffering. And I am equally curious about their accusers—mainly professional colleagues, but sometimes patients' family members, who are chagrined by what they perceive to be modern medicine's sprint to judge the quality of people's lives, as well as an unseemly haste to relieve suffering at the cost of accelerating death.

No Good Deed

1

The Police Came Knocking

At eight o'clock on a blustery Wednesday night in January 2001, two detectives from the state police knocked sharply on the door of Amy Gleason's home. Amy had just returned from a twelve-hour shift at the hospital and was wearily and mechanically preparing her supper. The two men were impressively large, and the atmosphere quickly turned confrontational as they introduced themselves.

"The state police were at my front door—so, stupidly, I invited them in," Amy explained to me several years after that fateful day. "Since then, I've learned a hundred things about police detectives. First of all, don't be in a hurry to invite them into your house, and second, you shouldn't go downtown with them when they ask you."

After walking into Amy Gleason's cheerful, cranberry-colored kitchen with its bleached maple cabinets, the two men politely refused to sit down. They quickly glanced around the room before turning their attention back to Amy.

"There's been a death," one of them somberly announced, pausing and leaving a strange emptiness in the air after the word *death*. The only thing Amy could think was that it must have been someone close to her. Her husband was at work, and she immediately thought of him.

"What happened to my husband?" she whimpered.

A look of confusion stretched across their faces, and they stared at Amy as if she had two heads.

"What do you mean, what happened to your husband?" they mumbled back.

"Well, you just told me that somebody died!" she choked out.

"It wasn't your husband," they said.

Instant relief, but suddenly her pulse began to race as she thought about her elderly father.

"Did my father die?" she gasped, wondering why else the state police would be at her house.

"No," they responded, continuing to be vague about why they were now standing in her kitchen. The strange dance of questions and answers went on for a while with Amy still unaware of what had brought these detectives to the house. Towering over her, they asked Amy if she worked at Baystate Medical Center. She confirmed that she was a nurse on Wesson 3, the hospital's renal unit. The detectives then asked if she knew someone named Rosemarie Doherty. Amy gazed at the men blankly for a moment as she tried to remember where she had heard the name.

"At first I had no idea who they were talking about," she later recalled when I spoke to her, "but then I realized that Rosemarie Doherty was the patient of another nurse, Kim Hoy. Honestly, I had had so little interaction with Mrs. Doherty that I hardly knew her name. I told them, 'Oh, yes, she passed away.'"

The two officers looked at her as though waiting for her to fill in the blank that they'd provided.

"Does that mean anything to you?" they finally asked.

Amy laughed kind of nervously at the question.

"Oh, that happens all the time."

Sitting back, years removed from that night, Amy contemplated those words.

"Something else I've learned is that when state police detectives are in your home, another thing not to say is that deaths happen all the time. That is a very, very bad thing to say. But I still did not have a clue as to what they were talking about. Maybe I was just totally naive, but I had no idea they were going to be accusing me of murder."

As Amy was learning what not to say to police officers, her fellow nurse Kim Hoy remained completely ignorant of the furor that would soon envelop her own life. Only after the state police were completely finished interviewing Amy were they going to pay her a visit. During my own subsequent interviews when I asked Kim to tell me about Rosemarie Doherty, I was shocked to watch her usually animated, intimate, and charming self instantly transformed into that of a vulnerable and frightened young woman. My conversations with Kim on this subject were always an emotional roller coaster. Most of the time we smiled at each other as she bubbled along, and then all of a sudden she would metaphorically, or in some instances literally, grab hold of my arm and I would feel as if we were uncontrollably falling. Her account of Rosemarie Doherty's last hospitalization began with one of those horrible descents into despair.

"I called her Rosie," she said, waiting a beat to let the name sink in. "She was so sick. She had been on dialysis for a while, and was septic, had vascular disease, diabetes, and emphysema. She had a bedsore which we were packing with dressings that took up her whole butt. Both of her heels were necrotic. Basically, she was rotting. Yes, there is no other way to say it. She was rotting.

"I had taken care of Rosie off and on," Kim said. "During this particular admission to the renal unit, I was her primary nurse for the first part of the week. I had a couple of days off, and then came back for the weekend. There had been active discussion about her becoming a DNR [do not resuscitate]. Although Rosie had periods in which she was more awake and seemingly alert, she was never fully oriented or coherent. Rosie recognized her children during their visits, but otherwise she was not cognizant of what was going on around her. She was terribly uncomfortable, and it was absolutely awful. I returned on the weekend, and the decision had been made during the previous night for her to be DNR and to receive comfort measures only. She would also no longer be going for dialysis treatments."

It was apparent to me that Kim not only cared for this severely ill woman's physical problems but also cared for her emotionally. While some staff can't help distancing themselves when a patient is physiologically and cognitively deteriorating, Kim remained tightly connected and committed to Rosie. Kim could not have imagined that her motives and nursing practices would be questioned, and that the police were going to knock on her door, too.

The detectives' visits to Kim and Amy had been set in motion several weeks earlier. A nurse's aide, Olga Vasquez, had returned home convinced that she had witnessed a monstrous act—there was no doubt in her mind she had seen a woman being murdered.

For some time Olga had been feeling progressively more uncomfortable with the treatment provided to one of her patients, Rosemarie Doherty, who had been staying at Baystate Medical Center. To begin with, the family and the doctors decided to stop offering dialysis, which meant she was going to die within a few days. On top of that, the primary nurse was Kim Hoy—a nice enough young woman, but one whose sense of propriety and boundaries appeared increasingly questionable to Olga. It was bad enough that Kim was prone to plunking herself down in bed alongside patients and reading them stories, but in Rosemarie Doherty's case, Olga felt Kim's behavior had been especially egregious. The patient was obviously having difficulty breathing; however, every time Olga turned on her oxygen, Kim would rush in and rip off the tubes and deliver a stinging rebuke about how the oxygen was not helpful. Olga became further convinced something was awry when on one such occasion Kim threw the plastic nasal cannula administering the oxygen on the floor and kicked it to the side. A few minutes later, Olga believed she saw Kim surreptitiously pick it up and conceal it in her uniform pocket.

These incidents were bad, but the final straw was the morphine. Olga carefully observed Kim and the senior nurse from the renal unit, Amy Gleason, taking opiate medication from the mechanical dispenser on the nursing floor. It was obvious to Olga that they took more than had been prescribed, and she was appalled to watch Kim inject the excessive amount into Rosemarie's drowsy body. Earlier in the shift, Olga had been horrified to hear Kim whispering to Rosemarie, "It is all right if you go now. . . . You don't have to hang on any longer." Accordingly, it came as no surprise to the nurse's aide when the poor lady quietly expired.

The irony of the situation was that Olga genuinely liked Amy Gleason and Kim Hoy. She had worked with them for a number of years and felt especially close to Kim, as they each had young children whose shenanigans were always an entertaining subject of conversation. Nonetheless, Olga was convinced she had glimpsed a dreadful and otherwise hidden aspect to Kim Hoy. Olga's faith and her personal beliefs left no room to ignore a crime, and she told her husband what she had seen. They promptly went to their lawyer and then to the district attorney's office. These visits took place at a moment in the winter of 2001, when many people in western Massachusetts were deeply preoccupied with the trial of Kristen Gilbert, a nurse from the Northampton Veterans Affairs Medical Center who had been accused of being a serial killer.

In the United States, it is not unusual at some point for doctors to shift from curative care, which focuses on healing people, to palliative care with its emphasis on the relief of symptoms and discomfort of dying. But within this seemingly simple change lie a host of issues with medical, political, and religious ramifications. There is also no clean, impersonal, and easy way to effect this change—there is no computer that a doctor simply turns off, no magic switch that gets thrown, no timer that runs out. It usually falls upon a nurse to go to the bedside and carry out a series of actions and complicated communications that allow the patient to die in as comfortable a manner as possible.

I am a physician specializing in palliative medicine and end-of-life issues. I have given presentations on these subjects at conferences around the world, but it was not until I came face-to-face with Amy Gleason, Kim Hoy, and Olga Vasquez that I completely understood the passions and the stakes of this work. Like Amy, Kim, and Olga, I work at Baystate Medical Center with patients who have kidney failure. Like these nurses, I have seen patients struggle to live when there is no hope of recovery.

However, unlike Amy, Kim, and Olga, I have been relatively insulated from the act of dying, having personally witnessed only a couple of deaths. The same can be said for many doctors, revealing the largely ignored truth that physicians are rarely present when people die. Instead, it is nursing staff who are frontline combatants when death arrives. Amy and her colleagues at Baystate Medical

Center have each ministered to scores, if not hundreds, of people as they took their final breaths. It may seem like hyperbole, but I believe really good nurses inherit the Crimean War legacy of Florence Nightingale, dispensing mercy at every battle. During my interviews with these and other nurses from my hospital, I came to marvel not only at their firsthand knowledge but also at their honesty and mordant sense of the ridiculous that allows them to adapt and even flourish in our pain-filled setting. Nurses see the suffering, the struggles to be cured, the families' anguish, and the denial, resignation, or acceptance when death is inevitable. Pulling the plug—a rather inelegant phrase—is a complex process that forces nurses to draw upon their empathy, personal convictions, religious beliefs, and professional training.

Not only are nurses often present at the terminus of life, but they invariably spend more time with dying patients than do doctors, and this places them at risk for attack when something goes wrong or gives the appearance of having gone wrong. Accusations against all medical professionals are serious, but with nurses they are exquisitely personal.

Still what happened between Amy, Kim, and Olga was not merely an accusation of malpractice or oversight. It was not about faulty clinical treatment or professionals falling asleep at the wheel. It was a manifestation of a far more serious matter, and one that repeatedly surfaces and informs conversations and debates all across America. If you haven't already participated in a decision to withhold or withdraw treatment from a loved one, sooner or later you are going to be faced with this dilemma. If you haven't already figured out your personal degree of comfort or discomfort over easing terminal symptoms at the expense of shortening people's lives, you are inevitably going to be forced to make this discovery.

Having conducted several research studies and written numerous academic articles, I thought that I knew a great deal about ending the life support of dialysis and vigorously providing pain medicines. I thought that I had overcome my reticence as an American to think and talk about death. But I was totally unprepared when these three staff members described Rosemarie Doherty's last hospitalization and how their disagreement over her care led to a formal allegation and investigation of murder. Like most people at Baystate, I had never previously heard of this highly confidential and largely secret incident. I didn't have a clue that such cases were also occurring in other hospitals around the country and around the world. And I didn't know anything about the philosophical conflict that underlies each of these cases.

Since uncovering the Baystate nurses' story, I have been able to interview a number of other accused medical staff, including Sharon LaDuke, who was fired from her position as nursing director of a rural intensive care unit after being accused of euthanasia; Dr. Robert Weitzel, who was convicted of manslaughter and sentenced to fifteen years in prison; and Dr. Lloyd Stanley Naramore, who was found guilty of attempted murder and intentional and malicious second-degree murder, and sentenced to concurrent terms of five to twenty years in a maximum-security prison.

I have also talked to others—on both sides of this issue—who are working to further their beliefs and causes. I have sought to understand the cases in a nuanced manner and have approached disability activists, religious leaders, health care authorities, political scientists, and bioethicists. I have also paid considerable attention to the crucial role of law enforcement, which has its own reasons for seriously attending to accusations. In the end, I have weighed all these different perspectives and sought to understand the broader context for the confrontations that occur between the proponents and opponents of palliative medicine. And I have decided that it is absolutely crucial for me to communicate how these conflicts can tragically mangle the lives of some of our finest caregivers.

2

Humpty Dumpty

"When I use a word," Humpty Dumpty said in rather a scornful tone, "it means just what I choose it to mean—neither more nor less."

"The question is," said Alice, "whether you can make words mean different things."

Lewis Carroll in *Through the Looking-Glass*

"Terri wasn't dying. She was cognitively disabled. It was needless. It was senseless. There was no reason to do this to my sister."

These words were quietly spoken by Bobby Schindler Jr., the brother of Terri Schiavo, whose sensationalized case about the removal of feeding tubes generated headlines around the world. I met with Bobby in December 2007 as part of my quest to make sense of what happened to the three nurses from my hospital—Amy Gleason, Olga Vasquez, and Kim Hoy—because I thought his perspective might be illuminating.

I found Bobby in Toronto, where he was the keynote speaker

at an anti-euthanasia symposium. Accompanied by his longtime friend, Brother Paul O'Donnell from the Franciscan Brothers of Peace, Bobby explained to me that Terri was a cognitively disabled woman who was neither terminally ill nor even calamitously sick. The two men nodded in unison and agreed that she died from "euthanasia by omission."

We were sitting in the lobby of the Toronto airport hotel, and there was a hint of apprehension as a snowstorm outside began to strengthen and people wondered whether they ought to secure earlier flights and get out of town.

"Life doesn't go back to normal after this," O'Donnell remarked. "There is a battle going on between the culture of life and the culture of death, and God has called upon our community to represent the culture of life."

Bobby Schindler was patiently sitting across from me, and he appeared unconcerned with the meteorological conditions. He explained, "It seems to be the premise of the other side that the acceptable alternative to human suffering is to kill . . . and I just don't go for that. I don't buy into the whole premise that killing is an acceptable alternative answer for someone who is suffering—whether emotionally or physically.

"For those who believe in the whole autonomy thing—that we should be able to decide the manner and place of our death—I don't think it is for man to really decide when our deaths should occur. Obviously, I believe that we are all made in the image of God, we are children of God, and He is the one who decides when we should leave this earth. It doesn't change if we become disabled.

"I look at it as a deliberate killing of a cognitively disabled person who had a family that was more than willing to care for her. I wholeheartedly believe that my sister was killed—killed by the system, killed by [her husband] Michael Schiavo, killed by whomever. She was deliberately and purposely killed."

When you converse with Bobby Schindler, it is obvious that his passion is sincere, as is his devotion to family, religion, and society's most vulnerable members. Earlier in the day, Bobby's formal address before hundreds of participants assembled at this Canadian conference had received an enthusiastic standing ovation, but whether or not you agree with the specifics of his point of view, the man's grief remains painfully raw and his argument is effectively delivered. His position is shared by millions of people from around the world, including the civil libertarian Nat Hentoff, who called Terri Schiavo's death "the longest public execution in American history." Similar beliefs are held by former president George W. Bush, who signed into law a resolution to prevent the feeding tube from being withdrawn, and later declared, "I urge all those who honor Terri Schiavo to continue to work to build a culture of life where all Americans are welcomed and valued and protected."

Although there are notable differences between the case involving the Baystate nurses and that of Terri Schiavo, Bobby Schindler mainly focused his comments to me on what he perceived to be their commonalities. In particular, when Bobby spoke about "the other side" and the people "who believe in the whole autonomy thing," he was talking about hospice and palliative medicine. This is the name of a new subspecialty that includes medical staff, like me, dedicated to the prevention and relief of suffering during catastrophic and life-limiting illness (throughout this book the terms *palliative medicine* and *palliative care* will be used interchangeably).

The word *palliative* first appeared in 1425, and though it was never commonly used, it was resurrected in the middle of the twentieth century to describe a medical approach that rapidly spread from Britain to the rest of the world. Palliative medicine has been proliferating in the United States ever since, progressing to the point that consultation services have been established in most hospitals with more than a fifty-bed capacity and a variety of hospices are available in every community.

Palliative medicine practitioners confidently work under the assumption that over the past three decades a quiet revolution has occurred. They believe that while it was previously considered improper to openly discuss death with patients and families, nowadays physicians and nurses should be required to have such conversations. They believe that our society approves of efforts to limit and curtail suffering—even if such efforts mean that dying will be accelerated. They believe that deaths following administration of analgesic medications and decisions to withhold or withdraw life-sustaining treatments are distinctly different from those that result from euthanasia or murder. They are confident that the procedures employed in palliative care are both ethical and legal.

According to the World Health Organization, palliative care is "the active total care of patients whose disease is not responsive to curative treatment." Therapies for these patients may include vigorous attention to symptoms such as nausea and pain, the provision of services at home or in nursing facilities, and psychological counseling for them or their families. Over the past decade, widespread acceptance of palliative medicine has contributed to a dramatic change in how people die. Even in America's intensive care units—our country's most medically aggressive settings—more than three-quarters of an estimated 400,000 deaths are now preceded by treatment limitation decisions. In the few short years between 2000 and 2004, the percentage of Medicare decedents enrolled in hospice care increased by 50 percent. In 2005, 1.2 million Americans died while making use of hospice services.

According to my friend Dr. Steven Miles, a professor of medicine at the Center for Bioethics at the University of Minnesota, fully 85 percent—or approximately 2 million—of the 2.4 million deaths occurring annually in the United States medical system are preceded by a structured decision to limit life-sustaining treatment. These are astonishing numbers. Over the thirty years that I have been practicing medicine, this represents a fundamental change—one that should have resulted in major repercussions.

If the practitioners of palliative medicine are on one side of a debate, then their opponents are people such as Bobby Schindler. For a substantial period of time I was unaware of the power of this other side and ignorant of the negative reactions that palliative medicine elicited from some segments of the general public. It was not until Amy and her colleagues began recounting to me the heretofore hushed-up details of the circumstances surrounding Rosemarie Doherty's death at Baystate, and not until the media became completely frenzied over Terri Schiavo's case, that I began to appreciate how situations like these are part of a direct societal reaction to the practice of palliative medicine. This was only reinforced by the tumult surrounding President Obama's health care reform efforts.

The Schindler-Schiavo family feud is especially instructive because it spotlights specific palliative care practices, such as the withdrawal and withholding of artificial nutrition and hydration (feeding tubes), and the proliferation of advance care directives (living wills, health care proxies) that are tilted toward helping individuals avoid the initiation or extended use of life-extending therapies (ventilators, dialysis, etc.). During and after Terri's death, the Schindler family has been instrumental in reinvigorating what Wesley Smith, an astute author and attorney, calls "a powerful and diverse strange political bedfellow coalition," consisting of pro-life groups, the religious right, a few outspoken medical professionals, activists for the poor, segments of the disability community, and elements within the Catholic Church. Although the coalition lacks an agreed-upon name, it is now a vigorous international movement that gathers under a banner proclaiming the paramount importance of the sanctity of life.

Activists in this coalition claim that organized medicine has become seduced by death, and some of them—but certainly not all—are steadfastly opposed to every medical decision that accelerates dying, including the withdrawal or withholding of medical technologies, vigorous use of narcotics, and so forth. For many of the religiously motivated members of the coalition—and they constitute the majority—the term *pro-life* has now been expanded beyond abortion to encompass the terminus of life and all practices that foreshorten human existence.

Back in 2005, the Schindlers and the coalition proved their ability to influence the governor and legislature of the state of Florida, Congress, and the presidency—even if they could not ultimately prevent Terri Schiavo's feeding tube from finally being removed. One year after Terri's death, the efforts of the coalition led twenty-three states to consider fifty-one separate legislative measures to change or nullify the use of advance directives. According to the coalition's view, these directives are objectionable not only because they make it easier for people to avoid lifesaving therapies but also because they frequently become a means for depressed people to end their lives and thereby passively commit suicide. A year later, twenty-six of the proposed bills were still active and Louisiana had changed its advance directive law, making it more difficult for patients to express preferences to terminate life support treatments.

The coalition is likely to have more success with three dozen so-called conscience bills pending in eighteen states that offer protection to health care workers who refuse to provide palliative care and thereby accelerate patient deaths. The conscience bills began as statutes to protect health workers from retaliation if they refused to perform abortions, but they evolved into laws allowing pharmacists and pharmacies to refuse to fill prescriptions for birth control or the morning-after pill. Now they are being extended to include medical professionals opposed to withholding or withdrawing artificial nutrition and hydration or performing other acts that may hasten death. State legislatures are building on a foundation of several existing federal laws that protect the conscience rights of health care workers, about which former Health and Human Services secretary Michael Leavitt has said, "Freedom of conscience is not to be surrendered upon issuance of a medical degree. This nation was built on a foundation of free speech. The first principle of free speech is protected conscience."

The coalition has been especially proud of having defeated recent efforts on the part of Vermont and California to legalize physician-assisted dying protocols; successful campaigns have similarly taken place in the United Kingdom and in Canada. While the coalition may be fragmented on some issues, it is united by unshakable antipathy to physician-assisted suicide. Consequently, the coalition is deeply disturbed that in the 2008 election, Washington's Death with Dignity referendum passed by a margin of 59 percent to 41 percent and Washington thereby became the second state after Oregon to offer terminally ill people the option of physician-assisted deaths. This was rapidly followed by a Montana judge endorsing a right to assisted suicide for a man with terminal cancer, raising the possibility of the practice becoming accepted in yet a third state.

The bioethicist Tom Koch has written in support of the coalition's efforts, "The unprecedented political involvement . . . from this perspective [is] wholly appropriate. Where else in a democracy do citizens turn when they believe current policy and law are inappropriate, prejudicial, and unethical? When courts cannot offer redress the logical next step is to seek legislation that will alter the laws in a manner that permits future judicial support."

To me, the Baystate nurse's aide Olga Vasquez has come to symbolize the views and suspicions of millions of people like Bobby Schindler and the sanctity of life coalition, who believe only God should make life and death determinations. According to this view the role of medicine is straightforward: combat disease and prolong patient lives to the utmost. As will become apparent, Olga's colleagues Amy Gleason and Kim Hoy see things quite differently; to them, the amelioration of suffering trumps the artificial extension of life. They maintain it is both logical and compassionate to withhold or withdraw medical therapies while providing vigorous symptom management, even if these actions result in people dying sooner than they would otherwise.

Amy and Kim believe their situation illustrates Clare Boothe Luce's aphorism "No good deed goes unpunished." Olga and the district attorney of the Commonwealth of Massachusetts would argue that the two nurses did not perform a good deed: the nurse's aide and the prosecutors affirm that they were misguided killers.

Terri Schiavo died on March 31, 2005. Since then the American public has remained largely unaware that the case's underlying conflicts continue to be actively contested. For that matter, most health care providers—American and foreign—do not realize a backlash is occurring against palliative medicine and its underlying philosophy. I have come to think of the Schiavo case as palliative medicine's Humpty Dumpty—irrevocably shattering the illusion that there is a societal consensus on end-of-life care. As Lewis Carroll declaimed in *Through the Looking-Glass*, we are now grappling over the use and meaning of highly charged words—we are struggling in the context of catastrophic and terminal illnesses to find the possible relevance of words such as *killing* and *murder*.

As the details of the Baystate disagreement are made clear, most people who have experienced the American medical system either as patients, family members, or care providers will be stunned to learn that this dispute resulted in a full-fledged criminal investigation. While doctors and nurses are always sensitive to the possibility that conflicts over how they practice medicine can result in civil litigation—malpractice—they remain entirely unappreciative of the fact that what they perceive to be ordinary clinical care can trigger criminal proceedings, including murder and homicide accusations. Most medical professionals are not aware that, without any forewarning, they may find themselves defending their actions and facing the possibility of bankruptcy, imprisonment, and the total obliteration of professional careers.

Some of my palliative care colleagues might have preferred that I not publicize this phenomenon, fearing a potentially chilling effect on the medical specialty. But as I learned about all that had gone on at my hospital, I felt compelled to write about it for the same reason road signs are posted about moose, deer, or other animal crossings—the notices may not always prevent crashes, but they at least heighten awareness and give us a slightly better chance of avoiding collisions. Ultimately, I am convinced that recognition of stories like that of Amy and Kim is the key to helping our society form a more rational response, policy, and practice. For too long, we have all (and I include myself in this) remained ignorant of the philosophical debate that is angrily taking place in our hospitals. It's time to open our eyes.

3

Setting the Scene

Amy Gleason is a forceful, powerful woman who laughs easily and is not inclined to complain. To my eyes she is pretty, but not of the delicate variety. She fulfills my image of how a serious golfer should appear—strong and solid.

"I look pretty darn average," she told me the first time we met. "I have a mop of thick hair in no particular style. I'm five foot ten. I don't wear makeup, and I don't fuss with myself. I'm just pretty plain and ordinary."

Amy said this as she and I were sitting across from each other at a restaurant, waiting for our food. As I eventually did with each of the nurses in this case, I had invited Amy to join me at a local restaurant because I wanted the conversations to take place in a more relaxed environment outside of the hospital.

Amy was born and raised in the western Massachusetts city of Northampton. "We had a very simple life," she explained. "We grew up on a street called Dewey Court, walked to grammar school, middle school, and high school. My grandparents lived with us, and my grandmother spoiled me, because I was the only girl. My mother never worked. She hung out and went to all of our games. We all played sports. She was one of those cheerleader moms who would run from my game to one of my brothers' games. We played outside like kids used to do, and came home when the streetlights turned on. It was a nice life in a very small New England town—certainly nothing special or fancy."

Amy smiled contentedly at me while recalling numerous bicycle trips to visit her cousins on the grounds of the town's lunatic asylum. Her uncle—their father—was director of the enormous and now shuttered Northampton State Hospital. Amy's favorite childhood memories involved running around with these cousins surrounded by acres of trees, fields, and the occasional crazy person.

When I asked Amy why she chose the profession of nursing, she answered in fits and pauses. "In my childhood, I witnessed a number of family members dying," she told me. "But I don't know if that pushed me into it. I took care of my grandmother when she was ill. She moved into our home when she had breast cancer with metastasis to the brain. Nursing is just something I was always going to do. All my friends were going to be schoolteachers and I was going to be a nurse."

Immediately following the death of her grandmother, it became apparent that Amy's mother had been denying or at least ignoring her own breast cancer. Amy was sixteen years old when her mother's disease was identified. I asked if Amy cared for her, and with a chuckle she good-naturedly barked, "I had two brothers and a father—need I say more? Yes, I took care of her for the next two years. Males aren't so good at those things."

When Amy was eighteen she started nursing school. "That first semester, I would get into my stupid little pinafore uniform every day and come downstairs. My mother would sleep on the couch because she had metastasis to the liver and had a huge swollen belly.

She would lift her head up and say to me, 'Oh, do you have to go school today?'

"I would think, 'I don't know why I'm dressed in this stupid outfit. Of course I don't have to go to school!' And I would go upstairs and change, and then spend the entire day with her. My professors were all very understanding, but it just got to a point where I was missing more nursing school than I was attending."

Amy dropped out. After her mother's death, she entered another program and eventually graduated with an associate's degree. Despite repeated entreaties from Baystate's administrators, Amy refused to pursue a bachelor's or other advanced degrees because she was content with the kind of nursing she provided and the kind of life she had created. Amy was not kidding when she told me, "Study? Golf is what I want to do in my spare time—not study."

After graduation, Amy left her home and moved to Boston. Her first nursing positions were at a long-term care facility and then at Massachusetts General Hospital. Fifteen years ago, when a nursing position opened up at Baystate on the renal unit, she packed up her things and returned home to western Massachusetts.

Now in her early forties, Amy was married for the second time and she absolutely glowed with quiet satisfaction when alluding to life with her current spouse. Unlike Kim Hoy, she never wanted children of her own and appeared quietly pleased that her stepchildren were adults. Despite having grown up in Northampton, Amy and her husband could not or did not want to pay its relatively inflated real estate prices. After they sold their condominium, she held her breath and bought a home in the nearby industrial city of Holyoke. She continued to frequently visit her father in Northampton but informed me that the new house and community had worked out nicely. Life really seemed to be on track for Amy until the police arrived.

• • •

Kim Hoy is an attractive, petite woman in her early thirties who by her own description is "a little person with big eyes"—they are strikingly blue and the most prominent feature of her face. When I would spot her on the renal unit, her brown hair was generally tied back in an equestrian braid that bounced merrily with her every step.

"I have been on Wesson Three [the dialysis and transplant ward] since it opened," she explained to me while we were at a Springfield restaurant. "When I first started to work I was pregnant with my daughter, who is now six. So it has been about seven years."

I swallowed a mouthful of food and said, "Okay. Just to fill in the picture of your family, you have a six-year-old, a ten-year-old, and a fifteen-year-old?"

"And a four-year-old—that's my baby," she proudly concluded.

We had briefly touched on this subject while waiting to be seated, and I had learned that her boyfriend was moving into her home along with his child from a previous marriage. I added, "Plus you are soon to have a five-year-old stepson?"

Kim must have recognized my bemusement at her situation (since I was quite content, if not overwhelmed, with the responsibilities of having two adolescent sons), and she smilingly said, "You're laughing at me."

"You are obviously a glutton for punishment," I responded.

"I cannot have any more kids myself," she explained with a chuckle. "So he is bringing up the rear."

Kim obviously adored children, and she went on to say, "Yeah, I never had much of a childhood, so I always wanted to have the pleasure of watching kids be kids. It's totally nice."

"How did you get drawn into the dialysis world? How did you decide on nursing?" I asked, switching back to the matter at hand.

"Well, there is the standard answer—I became a nurse because I wanted to help people. Actually, it never occurred to me that I would do anything else. I cannot ever remember wanting to do anything else. But beyond the standard answer is also the truth that I grew up in a really shitty way." Kim immediately apologized for swearing in front of me.

"You know," she went on, "it was really ugly, and awful, and downright crappy." She again looked remorsefully at me for her choice of language. "My brother died when I was very young. He was seventeen, and he was murdered—stabbed to death. He was a drinker and a partygoer. So when the person who stabbed him dropped him off on the front porch of his apartment building, several people stepped over his dying body because they assumed he was just drunk again. He got to lay there cold and alone, and he died while people literally walked over him. It puts things into perspective. I was thirteen when he died. I figured there ought to be something I can do to help people in this world.

"Meantime, I did what every kid who comes from a nutsy home does—I dropped out of school, used drugs, and went crazy for a couple of years. Then I grew up and matured a little bit, and I thought, 'This is not helping!' It took me longer to go to college, because I had to make up all the stuff that I never did when I dropped out of high school. But, I accomplished the work and graduated from the Baystate Medical Center School of Nursing. I got married, had babies, and now I try not to let people die alone."

Olga Vasquez is a nursing assistant—which Baystate calls a technical associate or TA. Olga is a crucial member of the treatment team, but she is pretty low in the hospital's pecking order. Her roles include doing EKGs, obtaining blood samples, giving bed baths, setting up meal trays, and monitoring nutritional input and output. To the patients on the renal unit at Baystate, Olga is the attractive Latina woman who helps provide basic care.

When I asked around, everyone at the hospital began their account of Olga the same way—they would tell me she was beautiful.

She was born and raised in Manhattan by parents who had come from Honduras. During a telephone conversation, she described herself to me as "forty-one years old, five foot four, with a lot of Indian features, a pretty fine nose, full lips, and tan, clear skin." I quickly learned that Olga was married and the mother of two children. Like the other nurses I interviewed, she experienced family illness and death in adolescence. In her case, she dropped out of college to care for her father, who died when she was nineteen years old. More recently, Olga had returned to school at night in pursuit of further education and a degree as a licensed practical nurse.

In our hospital, Olga elicited markedly polarized reactions from the staff: some clearly admired her, while others viewed her suspiciously and with a bit of apprehension. It has been impossible for me to determine how much of the latter opinion is retrospectively based on her role as the accuser. A nursing supervisor told me that she was absolutely adored by many of her patients and their families.

During my telephone contacts with Olga, she was consistently pleasant and sometimes emotionally effusive; however, she was impossible to pin down for a face-to-face interview. Olga was surrounded by family members who worried that she would not receive a sympathetic hearing from me and that her future job prospects would be damaged by negative publicity. Comfortable with her beliefs and actions, Olga was not especially worried about unfavorable consequences. I gave her the option of employing an alias in this book, and after inquiring and learning that Kim and Amy had chosen to use their true names, she initially refused my offer. She had a story to tell, and she was not about to shy away from it.

Although accusations of murder have taken place throughout the country and around the world, it is significant that this case occurred at Baystate Medical Center in western Massachusetts. With three nurses as different as Amy, Kim, and Olga, it's important to understand that their diversity mirrors that of the towns and cities around them. Most people reflexively associate Massachusetts with its bustling easternmost metropolis of Boston, and few are familiar with the comparatively sedate communities situated in the west, such as Northampton and Springfield. This part of the state is dotted with small urban areas amidst verdant hills and is influenced by the high concentration of colleges and academic centers. The broad Connecticut River constitutes the region's outstanding topographical feature, and its fertile floodplain has been compared to that of the Nile delta.

Northampton and nearby Springfield have consistently attracted unconventional individuals. Northampton has just celebrated the 350th anniversary of being established by a hardy group of Puritans, who demonstrated an adventurous streak in their willingness to leave Europe for the New World and their purchase of land from local Indian tribes for 100 fathoms of wampum, ten coats, and assorted bangles. In 1727 Jonathan Edwards was ordained a minister at his grandfather's church on Main Street, and he went on to preach the celebrated sermon "Sinners in the Hands of an Angry God." Edwards led a religious revival that swept through the country, and his Great Awakening is considered by some authorities to have been a precursor of the Revolutionary War. In the nineteenth century, before the outbreak of the Civil War, Northampton became a stop on the Underground Railroad. Sojourner Truth, a former slave, major abolitionist spokesperson, and feminist activist, was an active member in one of its several utopian communities. Sylvester Graham, remembered now solely for the cracker that bears his name, was an early advocate of health foods; his original house on Pleasant Street is presently a charming restaurant. On account of its vibrant lesbian and gay population the town has been the subject of a *20/20* segment and a cover article in the *National Enquirer*. An item in the *Los Angeles Times* described Northampton as being "a kind of lesbian Ellis Island. 'All lesbians pass through here at least once,' the common wisdom holds." Indeed, its mayor, several police officers, and a number of leading business figures have not been shy about publicly declaring their homosexuality. Northampton's current population is a mixture of Yankees, Polish farmers, deinstitution-alized mental patients and alcoholics, well-heeled and brightly tat-tooed Smith College students, proud Hispanics, and highly educated professionals. As depicted in Pulitzer Prize–winning author Tracy Kidder's *Home Town*, these disparate groups somehow manage to coexist in a politically progressive and remarkably artsy small city.

Although Northampton has its own small community hospi-tal, many people seek more intensive and high-tech care at Bay-state Medical Center within the nearby city of Springfield. Over time I have grown extremely fond of my hospital and the "City of Homes." Springfield still showcases some extraordinarily beautiful Victorian houses, although the industrial areas largely overshadow their delicate elegance. While I prefer to live in the more bohemian Northampton, Springfield has had its fair share of accomplished and imaginative residents. They include draftsman Milton Bradley, who manufactured the first mass-produced parlor game; Everett Barney, who invented modern ice skates; and Charles and Frank Duryea, who built the first automobile assembly plant. Theodore Geisel, better known as Dr. Seuss, was yet another native son, and he is hon-ored by several bronze statues that are prominently perched in front of the Springfield Museums at the Quadrangle. The once-popular Indian motorcycle was produced in this city. Humble resident and taxidermist Clarence Birdseye became the father of modern frozen food, and Professor James Naismith invented the game of basketball here, too.

Baystate Health is the chief tertiary care center in this region, and it shares some of the independent and even quirky qualities of western Massachusetts. Created out of the merger in the mid-1970s of Springfield Hospital Medical Center, Wesson Women's Hospital, Wesson Memorial Hospital, and the more recent additions of Franklin Medical Center and Mary Lane Hospital, Baystate is a modern amalgam of these small Victorian and Edwardian community hospitals.

In 1951 at the Springfield Hospital Medical Center, Dr. James Scola, a general surgeon, performed one of the first human kidney transplants in the United States. He made use of a healthy kidney removed from a patient dying of cancer. The transplant was rejected in a few days, but Dr. Scola's patient survived almost six weeks. Nowadays, Amy, Kim, Olga, and I work with Baystate renal transplant and dialysis patients who survive ten, twenty, or even thirty years. The nurses are present when their patients receive new kidneys or begin dialysis treatments, and they care for them throughout any and all complications.

Patients whose lives are extended with dialysis constitute the perfect lens to observe many of our modern society's complicated ethical and clinical issues. Unlike people maintained with ventilators, they generally carry on with their lives and relationships outside of the hospital setting, their thinking and consciousness is not clouded by medications, they are able to freely converse about the requisite compromises of treatment, and these individuals can determine when they have endured enough and want to stop.

The western Massachusetts population that seeks its medical care from Baystate tends not to frenetically migrate around; as a result, the hospital's medical staff, patients, and their families often remain intimately connected for decades. In this region there has been widespread acceptance of the tenets of palliative medicine, and I have consistently found the nine local dialysis facilities to be very willing to consider and implement its practices.

But one should not mistakenly believe that this philosophy of end-of-life care has been uniformly or unambivalently accepted by all patients and staff. Indeed, in a place with a population as varied as western Massachusetts, there really is not a consensus on these issues. With the diverse religious, ethnic, and socioeconomic backgrounds found in the region, the discussion threatens to grow more complex every year. Palliative medicine may be the current predominant philosophy, but a consensus will remain an elusive goal.

4

Rosie

"Rosemarie Doherty wasn't very old—perhaps in her sixties," Amy said, referring to the patient whose death resulted in the murder investigation at Baystate Medical Center. "She was not doing well on dialysis and had opted to stop. I was never her primary nurse, so I do not know all the circumstances around her death, but I know a conscious decision was made by her family to end the dialysis. Kim Hoy was taking care of her, and I think she was being followed by the medical house staff—the interns and residents."

Even though Amy and I worked on the same ward at Baystate and I am recognized for my expertise on dialysis discontinuation, until we sat down together for the interview I had never been privy to even the faintest rumors about the case. I was astonished to hear her story.

It was over a weekend, Amy recalled, when Rosemarie Doherty was actively dying.

"Rosemarie appeared to be suffering, and I remember Kim having a hard time because she did not have enough medications ordered. The pain medication was for a pathetically small amount of morphine, perhaps one milligram every four hours as needed. The lady was obviously struggling—I think it was her breathing, and she was anxious. It is a terrible thing to witness."

Amy paused to take a sip of water and collect herself as she remembered Rosemarie's pain. As she began to vividly reexperience the day, her narrative shifted between past and present tense. When I interviewed Kim, she did exactly the same thing. This behavior is typical for traumatized individuals who are trying to master what has happened.

Amy told me, "Kim keeps running out of the room to call the house staff to get more medication ordered, and of course they are reluctant because they think they will kill her, and God forbid you do that even though she has stopped dialysis!

"So they are hesitant to order more medication. I know Kim feels like her hands are tied. She has assured the patient and the family that we will keep her comfortable, and it is not happening. I remember it was time to administer Mrs. Doherty's morphine and Kim and I go to sign it out together according to hospital protocol. But Kim says she is expecting a call from the house staff to increase the dose, and she suggests that we sign out more in the hope that Mrs. Doherty will get the order for a higher dose. If they do not call in time, Kim says, we will just give the smaller amount.

"I go with her. Because Mrs. Doherty's is a partial dose of the standard units available in the hospital, we sign out the morphine together. When you give a partial dose, you have to witness it. I am her witness.

"Ideally in a perfect world, if I was your witness, you would draw it up in a syringe and then waste—squeeze into the sink—the excess drug in front of me. But, in practice, we never did exactly that. If you said you needed three milligrams of morphine, we might sign out a four-milligram vial, and I would trust that you were going to waste your one milligram and administer the three-milligram dose to the patient. I never questioned that you would not. You are a professional, and I would be confident that you were doing it."

Since this is largely a nursing matter and I am a physician, I had never hitherto heard about wastage. As I learned after talking to Amy, wasting is a hospital procedure that applies to controlled substances such as morphine, Valium, Demerol, Dilaudid, and even Lomotil (for diarrhea). When I later discussed the practice with several hospital administrators, they concurred that unobserved wastage by staff was common at our medical center—as well as at many, if not most, other clinical facilities around the country. The Baystate administrators assured me that this situation has now been rectified.

"So that night we each learned a huge lesson," Amy remarked. "We signed out the morphine together. Although I did not actually watch Kim waste the excess, I know she administered only what she was supposed to give. And then later, I believe, the orders were increased and she was able to properly medicate the woman and make her more comfortable.

"Kim and I still do not know everything that transpired, but apparently Olga Vasquez was deeply troubled about Rosemarie's case. At the time, Olga did not appear to have any issues. She never said anything to Kim or me about what was going on. She never asked, 'How much morphine did you use?' Nothing was said to indicate there was a problem. Kim wasted whatever we were supposed to waste together, administered the medication, and I think very soon afterward the woman passed away."

Amy looked simultaneously puzzled and indignant. "But apparently, as we would learn later, Olga Vasquez had a *major* issue with what occurred that night.

"At the time, I did not know anything about this. I found out the hard way. We had worked the weekend and I think Rosemarie Doherty died on Sunday. We worked, went home, and aside from the fact that a patient died, it was a typical day. You did not think anything unusual happened except, you know, Kim's patient died. I went back to the hospital on Wednesday, returned home, and that night the state police came to my door.

"After talking to me for some time, the police detectives said, 'There has been a complaint, and it is a suspicious death. We need you to come downtown to answer some questions.'

"I was flabbergasted and could only croak, 'You are kidding.'

' "No,' they replied, 'we need to talk to you about this.'

"And I said, 'Why? I mean, a lady died—people die all the time at the hospital—it is not unusual. She stopped dialysis, she chose to die, and she died. So why are we having this discussion?' But at that point the officers refused to go any further into the matter, and they demanded we go downtown."

Kim was the primary nurse for Rosemarie, and her account is similar but comes from a slightly more informed perspective.

"Rosie was a renal patient, but she was admitted under the general medical service and had medical house officers—that is, residents and interns—who were taking care of her. Unlike the nephrologists, these young and inexperienced doctors are unaccustomed to managing patients after the decision has been made to stop dialysis and to institute full palliative care. So when I came for the daily shift report, she had weird orders, including one milligram of morphine pushed intravenously every four hours as needed."

The problem, Kim said, is that when you give morphine intravenously, it is likely to wear off before the four-hour period is up. Slipping into her own version of Gertrude Stein, she explained to me that "pain is pain is pain is pain."

She went on to say, "If someone has given me the orders to deal with pain, then that is what I'm going to do. It does not mean 'Deal with it a little bit.' It does not mean 'Okay, let us make it not so bad.' It means giving sufficient morphine or another analgesic, and administering it frequently enough that the patient is no longer hurting.

"I needed to get changes made in Rosie's orders because they were not consistent with comfort measures. That is something I take very seriously. When you tell me a patient is to receive comfort measures, my sole function is to maintain comfort for that person—however I need to do it and with whatever I need to do for it. For instance, I don't want to see point-of-care or other unnecessary things ordered."

A point-of-care order is typical for the treatment of diabetes and involves using a lancet to draw blood every few hours, testing for sugar, and providing varying amounts of insulin. Kim's point was that when pain and symptom management is the principal goal, then it no longer makes any sense to inconvenience or to hurt dying people by sticking them with needles. Whether it is waking patients up in the middle of the night to give medications and take blood pressure readings or repeatedly drawing blood throughout the day to document laboratory findings, such things become nonsensical at this point in treatment.

"So I had an agenda," she recalled. "The first thing I did was to go into the room and assess her for pain—and she was in pain. I medicated her with the morphine push and then went straight out and called the intern. I asked the intern for changes in the morphine order and told her one milligram every four hours, as needed, was nothing. She had to give Rosie more. The intern was in her first few months of medical training and was hesitant about doing too much—like she would be killing somebody—but she did increase the dosage to two milligrams intravenous push every two hours, as needed. I asked her for an intravenous drip that I could titrate, but she was not too comfortable doing that, and I said, 'Okay, fine.' "

Kim was a relatively junior nurse, but she had a wealth of experience compared to the young doctor in dealing with such situations.

Hospital protocols required that she administer medications only as ordered by a physician, but she was not shy about making requests.

I asked Kim why she preferred Rosie to be on a morphine drip, in which the medication is continually infusing into the patient's vein. She clarified, "A morphine drip gives me the liberty, without having to call all the time, to either go up or come down on my rate of analgesia. If I walk into a room and see someone is really uncomfortable, often they are unable to tell me with words. If they cannot describe it to me, I look at their eyebrows. You can always see it in the eyebrows. If someone is having pain and is a little wifty, they might moan and move around, but more often than not their eyebrows will get a little . . . I don't know what you call it, but the eyebrows change, they move—they squint.

"You need to look for pain in different ways. If someone cannot tell you they are having pain, then watch the respiratory rate. Observe how they are moving in the bed. Do they sit for a second and then become restless? Is their blood pressure in the nineties and you gave them a bed bath and it is now 145? What is their pulse rate? Are you seeing changes? Because the most terrible thing that I can imagine is to have somebody lay there in pain and for me not to have looked closely enough to recognize it. There is nothing worse than for a person to be in pain and unable to communicate to me what they are feeling."

When Kim said those words, I could not help picturing her brother lying in the street and bleeding to death. The image was palpable.

"A drip would have given me more of a range," Kim went on. "If I come in and people are lying there comfortably, I am going to leave the drip rate right where it is. I do not like administering morphine through injections, because you give it and the patient gets really comfortable, but then the analgesia begins to wear off. They are going through what we call 'peak and valley' discomfort.

Even though you tell me I can give the injection every hour, unfortunately that is not my only patient. So if I get tied up over here with Mrs. Jones, then I cannot get back and am unavailable to give the morphine to Rosie. At least if she is on the intravenous drip, she is always getting something." (There are also newer patient-controlled analgesia pumps that deliver a constant rate of morphine but allow patients to supplement the dose themselves by pushing a button. Unfortunately, this was not an option for Rosie, whose consciousness was impaired.)

Kim proceeded to detail another practical reason why she would have preferred orders for a morphine drip, "Rosie was incontinent of stool. I am pretty sure she had C-Diff [a nasty and common infection of the chronically ill]. So one of the problems was she would stool, and it would be from head to toe. We would then have to move her, and wash her, and pull that packing out. It was very stressful. Her respiratory rate would rise, and she would begin to moan. It was terrible. With a drip, you can anticipate these events and can bump her one or two milligrams while you are cleaning things up, and then afterward you can turn it back down. It gives you a little more liberty."

As she spoke, Kim's hands constantly moved through the air to punctuate her comments. "I reviewed the orders with the intern— the dressing changes, point-of-care, lab draws, and all the other kind of stuff. Rosie was going to die—it was inevitable—so why change the dressing a million times? It made more sense to keep the dressing clean and change it as needed, rather than on a fixed schedule. I brought up the issue of administering oxygen—Rosie had emphysema—and I had her discontinue the oxygen order.

"Usually if you have emphysema," Kim told me, "at most you need really low-flow oxygen. Put too much oxygen on patients and it just makes them repeatedly gasp; their respiratory rate goes up higher and you have to deal with that. Did you ever try breathing thirty-two breaths a minute? You get tired mighty fast. It really sucks!"

A palliative medicine colleague supported this position, saying that Rosie's shortness of breath was not necessarily prompted by decreased oxygen and would not automatically have improved by administration of the gas. He told me that many people under those circumstances may end up feeling considerably worse if forced to receive oxygen, particularly through a closed mask, and they would likely have a better response to ordinary room air blown over the face with a bedside fan or an open-faced mask coupled with the use of opioids. While this is what Kim eventually offered, she was never able to effectively deal with Olga's distress at the removal of the oxygen, which for many patients, families, and staff is a potent symbol of medical care.

I asked Kim to estimate the number of her patients who had stopped dialysis or died for other reasons during hospitalizations, and she became visibly flustered. Kim's hands began to weave around and she hemmed and hawed. It seemed she could not reduce these nursing experiences to numbers; the intensity of her involvement with people kept the focus on them as individuals.

As I heard about Rosemarie Doherty, it became quite apparent how nursing care can easily be subject to interpretation or misinterpretation. But Kim was quick to point out that not all cases involving terminally ill people lead to finger-pointing and recriminations. There are stories that exemplify what happens when palliative care goes well, and as she said this, she became visibly more at ease. Inhaling deeply, she said, "Let me take a few minutes and also tell you about Barbara Dilanian and Benjamin Babcock."

5

The Wedding

"I was Barbara Dilanian's nurse when she had a kidney transplant after developing renal failure," Kim began. "She then ran into a lot of trouble with infections, had some rejection [of the transplant], and wound up getting cancer. Barbara was treated for the cancer, but after that her renal function never came back. She went into chronic rejection and ended up resuming dialysis. She was sick—really sick.

"Throughout it all, Barbara's husband, Ken, was wonderful. He was determined to help and comfort her. It didn't matter what happened—he was right by her side the whole time. He was always rubbing her head. I will never forget that as long as I live. Barbara said, 'Rub my head, honey. Rub my head.' And Ken would just rub her head. Even after chemotherapy, when her hair was gone, he rubbed the top of her head. That was her relaxation; it was her comfort zone."

As Kim spoke about Barbara, it was clear that the memory was painful in the way that all stories about dying patients are painful, yet there was something more behind it—a sense of peace that can easily be absent from conversations about death. I brought this up with Amy, who was the senior nurse and primary caregiver for Barbara Dilanian, and she drew a more complete picture of what was so special about this particular case.

"Barbara was a lovely lady. She was admitted several times, and during the last hospitalization she spent a chunk of time in the intensive care unit. I think the day we hit it off she had just been transferred from the ICU, and I told her husband, Ken, to go home and get some clothes and personal things so that she could look like a human being again.

"I remember saying, 'You know, Barbara, you kind of smell like an old goat.' And she did! She was all gross from being in the ICU—where they are more into saving lives than personal care. Well, Ken drove home, and shortly afterwards he came back with all new stuff—dressing gowns, pajamas, underwear, and so on. Yes, I told her she smelled like a goat, and from then on we hit it off!"

Amy looked carefully at me and tried to gauge my reaction to her goat comment. Frankly, I thought it was endearing, but it also said a lot about the kind of honesty that informs Amy's conversations and relationships with people.

"During that final admission," she continued, "Barbara was getting sicker and sicker. She went back on dialysis and hated every minute of it. My impression was that she knew because of the cancer she couldn't get another kidney for a transplant, and she just hated the way her life was going. She hated it. But she loved her husband, loved her children dearly, and loved everything else about her life—except that she was sick and dying. She always wanted to change that one piece. If she could only figure out how to get better and have her normal life again . . ."

Amy's voice trailed off momentarily and she took a deep breath.

"Eventually she pretty much realized it was not going to happen. She got progressively sicker, and she finally made herself a DNR [do-not-resuscitate order]. This was a pretty big move, because she initially thought that DNR meant we were going to brush her to the side. We explained that when the DNR order was in place she would still be treated and could continue or discontinue dialysis. That was acceptable to her.

"It became obvious that Barbara was not going to live much longer. I knew months before she died that she was not going to survive, but she did not accept it. I suspect she kept hoping she was going to get better and everything was going to be fine. In the end, she thought she would go home. Plus she had her daughter, and there was going to be the wedding."

But the reality as Amy saw it was never going to match Barbara's expectations. Barbara had been sick for far too long and had never really healed from her respiratory problems. Her lungs were in bad shape, she couldn't breathe without a lot of oxygen, and she couldn't do any physical therapy because she was so short of breath. Her life was incredibly restricted by her condition. All she could really do was lie in bed and think about how terrible it is to be on dialysis.

"We talked all of the time about her daughter's upcoming wedding, how it was going to take place on Cape Cod in August, and what a big thing it would be. Barbara kept saying, 'I have to get better, because August is coming. August is coming!' The wedding was always her goal. She was determined to see her daughter, Jane, get married."

While this hope was a strong motivator for Barbara, it became increasingly apparent that it was not going to become reality. It fell on Amy's shoulders to level with her patient.

"One day I said, 'Barbara, you are not going to get better, and there will be no August.' I know she was surprised."

That same day, Barbara's daughter, Jane, was also visiting, and she had been sitting on the bed in her mother's room, planning the wedding and talking about her gown. As Amy tells it, she waited for the right moment and pulled Barbara's daughter out into the hall.

"I think your mom is going to die really soon—within the next couple of days," Amy said. Understandably, the initial shock was difficult, and it took Jane a few minutes to compose herself. Like Barbara, Jane and her whole family had turned the wedding into a focus for all their hope. Hearing Amy's words now contradicted everything that they had wanted to believe for weeks and months.

After Jane got over the initial shock, Amy asked her, "How do you feel about getting married now?"

Jane nearly fell over. "But—But I'm getting married in August," she stammered.

"Well, your mom's plan is to see you get married, and that is all she wants to do. So why don't we try to get you married?"

"When?"

"Tonight," Amy replied, amazed at the role she had suddenly assumed. Amy hadn't even planned her own weddings. Her first had been coordinated entirely by her father, while for her second she and her husband had simply run off and gotten married, no planning required. Now she was proposing to organize someone else's wedding—someone who was virtually a stranger.

"This is private." Jane demurred. "We don't need this."

But Amy is not a person who gives up easily, especially when it comes to a situation like this, where she knows the alternate endings and none of them are good. Amy continued talking to Jane for a while, trying to get her to see what Amy's years of experience told her. Finally Jane's fiancé, Allen, spoke up.

"Janie, I think this is something we need to do."

Jane looked at her fiancé, then turned back to Amy and nodded in assent. And just like that, the wedding preparations began.

Looking back years later and recounting the details of Barbara's story to me, Amy's voice was still enthused.

"It was great!" Amy recalled. "And it was great because our staff was wonderful. Suddenly everyone was pulling together. They were doing a lot of my work so that I could focus on the wedding. Otherwise it probably could not have happened, because you can't just ignore the care of seven other patients while planning a wedding.

"There were a few stumbling blocks. I couldn't get one of the Catholic chaplains to officiate. The Dilanians were not Catholic, they were Protestant—Congregationalist, I think—but that was not the problem. I called a priest to oversee the ceremony and explained that this would not be an official wedding. It was just a wedding so that Barbara could see something. It was not going to go down in the books as the actual wedding. Well, he refused, because it was going to be 'deceitful and lying.' Oh, for Christ's sake!

"The chaplain then showed up at the renal unit and tried to further explain his position, and I said, 'I don't really care. If you are not going to help me, get out of my way, because I have a lot to do right now.' I ended up calling the mayor's office to get a justice of the peace. That turned out to be a pretty good idea, and they were very helpful.

"In the meantime, Young Hee Kim, the dietitian, made arrangements for a meal of shrimp cocktail, filet mignon, and champagne. She went out and bought one of Barbara's favorite lemon cakes and decorated it with curlicues and stuff.

"As this event begins to take shape, Jane, Barbara's daughter, said to me, 'Oh, my God! I need my wedding gown, and it's in Boston at the dressmaker's!' Well, her fiancé, Allen, piled into his car and quickly drove off. Meanwhile, some of our staff went to the hospital gift shop and bought Barbara a new nightgown so that she would look presentable. We all helped Barbara with her makeup. While it is pretty hard to make a dying woman look nice, in short order she looked about as nice as you can look under the circumstances.

"The husband-to-be returned from Boston with a dress that was literally pinned together—little pins held the sleeves on and attached the body to the skirt. He came back with this fragile garment, and Jane grabbed her makeup and got dressed in her mom's room. Allen and her father drove back to the house to change into their suits.

"Jhun Ciano, the husband of one of my colleagues, does wedding photography, and he came over to videotape the whole thing. Jhun managed to record the standard goofy stuff, like Jane putting on her gown, the corsage being attached, and the ordinary pictures you see of a bride and her mother on their special day. So it was all on tape.

"A visitor to the renal unit noticed the commotion, and mentioned to me that she sang for her church, and asked whether she could participate. I said, 'Great! You're hired!' and she zipped home, changed into a fancy dress, and came back all decked out prepared to be the singer. We had the chapel decorated, and the night staff arrived. Somebody played the organ. I don't remember who did the readings, but the justice of the peace was absolutely wonderful.

"By this time Barbara was really, really bad. She was failing, and we were not even sure she was going to make it. We pushed Barbara and her bed into the chapel. It was one of those special beds and it was *enormous*. And we stood by during the ceremony, ready to inject her if she stopped breathing or her heart rate dropped. We had some drugs handy to give her, but fortunately we never had to use them. We decided not to drag the code cart [which contains the electrical paddles and other equipment for cardiac arrests] into the chapel, because that would have looked tacky."

According to Kim, the nurses tag-teamed together to keep Barbara alive.

"My biggest chore was taking additional patient assignments in order to free up Amy. Or at least that was my main role until the actual ceremony. Barbara was having a lot of problems with her blood pressure. She couldn't maintain a blood pressure to save her life—literally. We really wanted her to make it through the wedding, and because I am telemetry certified, I asked if it would be all right if I put her on the drug dopamine and titrated her blood pressure. We intended to keep her awake so that she could see everything that was going on. Accordingly, in the midst of all this wedding stuff, I had her hooked up to the cardiac machine. I was watching her blood pressure cycle and I was carefully adjusting the medications. As her blood pressure dropped down, I would go up a couple of clicks, and she would plateau a little bit."

It was that kind of back-and-forth team effort that managed to get Barbara through the ceremony, which took place on a Friday night.

"Our little patient lounge was set up as the reception area," Amy explained. "The dietitian got flowers—lots of bouquets—for Jane and the others. It was truly like a . . . it was like a normal wedding in every sense, except for the mother who was dying. There were flowers, there was wine, and there was everything you would expect at a reception.

"Jane marched in, they played the usual music—you know, 'Here comes the bride.' And throughout the service, we kept Barbara in the back in case we had to make a run for it. She was critically ill, but she kept saying over and over again how happy she was. It simply thrilled me to hear those words.

"Later I caught a glimpse of Jane and Allen sitting on Barbara's bed and holding her hand. It was totally worthwhile. Barbara was ecstatic. She had gotten to see her Janie get married. We played some songs, and the couple danced together. Allen is a wonderful husband."

I asked Amy to what extent Barbara was able to participate in the festivities following the service, and she replied, "One hundred percent! She was able to eat steak, even though she hadn't previously eaten in days. She had wine with her filet mignon, and finished things off with a big bite of wedding cake. Barbara was completely involved, and probably as mentally alert as she had been in a long time."

Amy grinned and added, "The food service department did some fancy footwork to prepare that dinner. I don't know about you, but I've never come across filet mignon in our hospital cafeteria."

I was curious and asked, "Was Barbara's dialysis discontinued?"

"Yes—Barbara chose to do that. I think it became clear that her medical situation was terrible. Cancer and the chronic lung problems were killing her, and dialysis was not going to help. I do not remember exactly when the decision was reached to stop the dialysis. She looked like she was dying for many weeks, but I think she was dialyzed up to the last day or so. If I remember correctly, she might have had dialysis on that Thursday. She really stayed with it almost right until the end. I know that she was never dialyzed again after the wedding."

Kim has developed strong opinions about coercing patients into undergoing dialysis.

"We are all going to croak—every last one of us," she told me. "Stopping or not starting treatment are almost never presented as options—and they should be. Some people are eighty-two years old and we are seriously considering beginning dialysis? What is the point? Where are we going with this? It does not always make sense for us to apply technology and interfere with what would otherwise be occurring naturally."

Although smiling, she was deadly serious when she said, "I have gotten more than one dirty look when a family member whines, 'But *dear*, you *have* to do the dialysis.' I look directly at the family member—child, spouse, or parent—and I tell them, 'This is not something they *have* to do, and they should certainly not *have* to do it because of you. If they'—and I point to the patient—'want to do it, that is fine. But you are not the one who has to be on the machine. You are not the one who has to feel what they feel. What you should be doing is supporting their decision—whether you like it or not. Frankly, whether you like it or not doesn't really matter.'

"On top of being sick and having all these big, life-altering choices ahead, the last thing anyone needs is to be browbeaten by their family into continuing to do something that they don't want. This is especially the case when we are dealing with older adults—people who in some cases have lived many, many years, and have been happy during their lives, and now they are accepting that their lives are going to end. Some families seem to forget that sometimes the way you spend the time that you have left is more important than simply being here and being alive. You know, anybody can just be alive. But what is important is how you are alive.

"Almost everything changes when you start dialysis. It is real easy for medical staff to announce, 'Okay, you are going to go to dialysis every Monday, Wednesday, and Friday. You need this surgical thing put in your arm. Oh, yeah, and now you are going to have to change your diet and do this and do that. By the way, you are also going to be mighty constipated and probably itch a lot.' It is real easy for families to say, 'We think you should be dialyzed.' What is hard to accept is that Dad or Mom is going to die, and I have to support their decision even though I don't like it. We get selfish. We get caught up in what is going to feel better for us and not what is going to feel better for them.

"Barbara Dilanian's family was not like that. It is probably because of how she raised them, and the type of person that she was herself. She was a giving, kind person. She wasn't selfish. In a lot of ways, she was selfless. And her kids got that from her. There was no browbeating, and there was no bartering or requests that she continue dialysis for another three months. Her family listened to what she said, and they respected her preference. They saw the stuff that she had been going through, and they knew that she was making the right choice for herself."

Kim observed, "The wedding was really kind of neat. Barbara got to see her daughter married, and had a nice little meal and even ate some food. Afterward, over the next few days her family visited, talked, and played tapes. They looked at pictures from when everyone was younger. They sang, remembered family vacations, and recalled good and bad times. They had an opportunity to remember all those years of life, sorrows, happiness, and old sweatshirts."

I asked Kim about the old sweatshirts comment, and she explained, "Just different stuff—little things in families that we all have. Like, the way the oldest son always has the stinky socks. He always gets picked on for that. In my house it is this noise I make with my nose, because I have bad sinuses. That will be the running joke for the rest of my life. There is always something about each family member that everybody picks up on and remembers—a nickname, a peculiarity.

"Barbara's family was able to go through all of that and say goodbye, and they were able to be comfortable that she was going to die. They understood what was happening to her. There was no surprise. They all knew exactly what was occurring and that it was taking place in a safe, controlled environment with people they knew. She had been in and out of the renal unit so frequently that they knew us. They knew us by face, and they knew us by name. When the family couldn't be present, they were confident that we were there. It was a completely secure place. It was really kind of neat.

"If Barbara got anxious, we could help deal with that. If she had pain, we could also help deal with that. Her last few days of life were not anguish, torment, or discomfort. Her family was protected from seeing her suffer. That has got to be the worst thing for a mother—to have your family watch you wilt away while there is nothing they can do about it. Well, it was not like that for them. They had us here to take care of all those things, and that is what we did. We provided the care, so that their last days with Barbara were not terrible memories of pain with everybody crying and sobbing."

Kim paused and looked at me as if assessing how I might respond to what she was about to say next.

"Bear with me, because this may sound weird," she finally said.

"The Dilanian family's experience was almost the opposite. It was more like when you are pregnant and so excited in those last few days before the baby comes. You sit in the rocking chair, and you rock, and you talk to your tummy, and you say, 'I am ready for you to come out now, and your crib is ready.' You refold all the little clothes ten times, and it is exciting, and it is fun, and you know it is a time that you will look back on and remember fondly. You are getting ready for this big event in your life, and you are prepared for when it happens. Okay, admittedly this was not something as joyous as a baby; however, there still was that preparedness, that sharing, and the agreement that we were all ready now. Everybody was okay with it. It was really amazing. I mean, it was sad—it was horribly sad when she died—and I would never take that away from her family. They still suffered; they still felt lonely and missed her, and all the stuff that comes with death. But imagine the peace of mind that was shared. How many people get that?"

Barbara's children, Jane and Ken, were interviewed by Laurie Bobskill from the *Springfield Union News*. Jane told her, "I think at first my mother was going along for everyone else. But somewhere it changed.

"My dad is not emotional, but when he saw me in my wedding dress . . . four nurses in matching scrubs for bridesmaids . . . when the soloist sang 'The Lord's Prayer'. . .

"It kept getting better. It was like the soap operas my mother and I used to watch together, like *Days of Our Lives*. It changed the worst days of our lives into something so happy.

"We feel this was the real ceremony, more meaningful and spiritual than the August one will be. We feel married now."

At the time of Barbara's death, her son, Ken, was a correspondent based in Rome for the *Philadelphia Inquirer*, and had recently accompanied the 173rd Airborne Brigade to northern Iraq. Although he missed the ceremony, Ken quickly flew to his mother's side in the hospital. In short order, the photographer dubbed in the music and completed the editing, and Ken was then able to watch the entire event on videotape.

Ken was quoted as saying, "[Afterward in my mother's hospital room], we watched home movies on the wall. We brought in beer and Chinese food."

Barbara's thirteen-year-old cat, Tabby, was permitted to come for visits.

"The first time, Tabby stayed in the carrier," Jane said. "The second time, Tabby slept on Mom's stomach for two hours. It meant so much to her."

Because Barbara would not live to hold grandchildren, one of the staff members had his cousin bring in her own baby. Amy remarked, "When someone is dying, rules can be broken."

As Barbara began to drift into sleep under the influence of a morphine infusion, Amy said, "The kids talked to her. I assured them she could still hear. Jane came and told her all about her funeral. She's a little spitfire. There's a lot of Barbara in that girl. She told [Barbara] that she had made their father buy the expensive casket."

During our interview, Amy told me, "Barbara's family was there when she died. The kids had been sleeping in the room, and I believe her husband was with her. The wedding was on a Friday, and she died early [Tuesday] morning. She kind of just drifted off, and that was the end. I do not think she had a terrible death. It was a fairly comfortable death. I think she died immediately before my shift started, because I remember coming in, she was there, and she was dead. Her children had time to accept it, and it was not a big scene. I think they were all grateful for the weekend.

"The attending nephrologist pronounced her dead. We then did the postmortem care. I think people probably do postmortem care differently. In Barbara's case it certainly felt different, because Barbara was so special. You give the person a last bath so that they are not sent off looking . . ."

Amy's voice trailed off as she recalled those final images and moments.

"I don't know how necessary that is, but it is what we do—a last bath. We are not supposed to put anything on them. I have actually been reprimanded for sending people to the morgue in johnnies and blankets, but I always put fresh johnnies on them and wrap them in a clean blanket. You are not supposed to do that because it is a waste of linen. But the thought of putting somebody naked in a plastic bag is really repulsive to me. So I do what I do."

Kim looked especially somber when she told me, "I was not in the hospital when Barbara died. I returned after a couple of days, and usually the first thing you ask when you come on your shift is, 'Did she go yet?' You walk into work, and you look up on the board and see if the name is still written there with all the names of the active patients. If her name is not there, you kind of already know before you ask, but then for some stupid reason you ask anyway. You want the confirmation. Well, I asked about Barbara, and she had passed away. I am thankful that the family was with her."

"I went to Barbara's funeral," Amy softly said. "It was probably the nicest funeral I have ever attended. It was in East Longmeadow at the Congregational Church. It was just what you would expect for Barbara. I am accustomed to going to Catholic funerals, where a priest stands at the front, mispronounces the deceased's name, and that is pretty much the end of that. At this funeral, people got up and spoke about how Barbara influenced their lives. It was extraordinarily nice and warm. It was a lovely send-off."

I asked Amy why she had made this wedding happen, and she replied, "I don't know. It just seemed like the thing to do. Once I started, lots of people helped.

"If I really have to make sense of what I did . . . Well, my mother wasn't around for either of my marriages. Especially my first one—the big one—which was a Catholic wedding with the gown and the whole bit . . . I think it is important for a daughter to have her mother at the wedding. So that's probably what prompted me to volunteer."

I enjoyed listening to Amy and didn't want her to stop. I wanted to hear more about her account and her observations, but there was one topic troubling me. I took her back to the very beginning and asked, "You were the one to tell Barbara she was dying. What about the doctor's role?"

Amy replied, "Barbara was in such denial! It was apparent to everybody else that she was dying, but not to Barbara. She just did not want to accept it. Everybody was sort of skirting around the issue. Nobody wanted to be direct with her. One day I just said, 'You are dying.'

"This sort of thing has become a nursing role. I am convinced we have made it one, because nobody else does it. We have taken it on as our role. If the doctors are not going to do it, then I think we have all decided that we need to do it. I guess some of the doctors are better at it than others. But I think they still wait too long. I mean, when you are ninety-eight percent dead and somebody then talks to you, it is kind of pointless. It has to be discussed much sooner. So we nurses do it all the time.

"It was difficult when I first started, but now I feel it is my job. I feel like I owe it to the patients, because they need to know. I think it is a disservice if we do not do it. I can understand younger nurses—newer nurses—not doing it, because it is certainly something that takes a level of comfort to perform. I don't think that if you are a new graduate, you can easily walk in and tell somebody they are dying. But I feel it is my duty and my obligation to do this with patients. If I think of myself or my family member lying there dying, I certainly hope somebody will tell me. I want my last few days to be very different. I think people can act differently if they know they are dying."

• • •

Barbara Dilanian taught third grade at Mountain View Elementary School and had been a special education aide at Longmeadow High School. Her husband was a vice president of a division of an insurance company, and they provided their two children with a solid, middle-class upbringing in the town of East Longmeadow, Massachusetts. Barbara's greatest source of pride was being a mother.

The hospital's medical records reveal that she was fifty-eight years old when she died and that the renal failure had begun seven years previously because of a common hereditary disorder, polycystic kidney disease. She received two cadaveric kidney transplants, and both were subsequently rejected. She was also diagnosed as having Hodgkin's lymphoma, and underwent several courses of chemotherapy and radiation therapy. All the while, her peritoneal dialysis treatment was complicated by staphylococcus peritonitis—this is a recurrent and painful infection of the gut.

Barbara's last admission to Baystate Medical Center came about because she had been having difficulty breathing. The hospital summary concluded that she succumbed to a respiratory infection called bronchiolitis obliterans—a form of pneumonia. She spent over a month in the intensive care unit before being transferred to the renal ward. Clots developed in the veins and arteries of her left leg, and she had complications from the anticoagulation medications. She chose to undergo vascular surgery to improve the circulation of her leg. During the hospitalization she developed a heart arrhythmia; it was treated and her heart rhythm went back to normal. Barbara had persistent respiratory difficulties, and after another month her condition began to further deteriorate. She decided to stop dialysis, and DNR orders were written.

The only allusion to the wedding occurs in a couple of brief notes. The morning following the ceremony, Barbara Dilanian was described as being very "bright," and her family was present in the room watching the wedding videotape. A nurse wrote in a progress note that this was an "emotional day for all."

Then after an episode of vomiting and aspiration, Barbara and her family requested that a morphine drip be started. She was quoted as saying to staff, "You promised"—referring to an earlier conversation about dying without pain and suffering.

The final nursing report noted that the patient was unresponsive and agonal breathing (a common pattern of respiration at the end of life that has nothing to do with agony or discomfort) was present. The family was noted to be in attendance and accepting of the situation.

At five-fifteen in the morning, an examination confirmed the absence of a pulse or respirations and that her pupils were nonreactive to light. Barbara Dilanian was declared dead.

Shortly after his mother's death, Ken Dilanian wrote an article that appeared in the *Springfield Union News*, entitled "Hospital Caregivers Show a Giant Heart." In the article he explained, "There wasn't time to get a marriage license or blood tests, so the wedding wasn't technically legal. The August ceremony will take place as scheduled.

"But for my sister and new brother-in-law—and for my mom—the Baystate nuptials were as real as anything could be. The event turned what could have been a day of despair into a day of hope and joy. And it buoyed us all, Mom included, through a weekend of togetherness during which we dwelled not on the unfairness of it all, but on reminiscences and love.

"Over the next three days in the hospital room, we fed Mom her favorite restaurant meals, watched home movies on her old eight-millimeter projector, and celebrated my dad's sixty-sixth birthday.

"On Tuesday morning, she died at peace as Jane and I slept beside her.

"We played the wedding video during the viewing at the funeral home. It tinged the sorrow with celebration."

6

Turning Points for
Planned Deaths

The ethicist Daniel Callahan has written with great delicacy about a "tolerable death," and although he had in mind people who were older than Barbara Dilanian, her death probably would meet his definition. According to Callahan, such a death takes place when one's life possibilities have largely been accomplished, moral obligations to those to whom one has responsibilities have been discharged, the act of dying does not offend others' sense or sensibility, and it is not marked by unbearable and degrading pain.

Callahan shies away from calling these "good deaths," yet I would say that in many regards, Barbara Dilanian had not just a tolerable death, but an exceedingly good death—an exceptionally good death. She had the kind of death that is sought as an ideal goal by modern palliative medicine practitioners; it was a contemporary or even futuristic death. Her life was extended with the assistance of devoted clinicians and advanced technologies in order to allow the achievement of a long desired family milestone. She died gracefully.

When I mentioned earlier that the way we die in the United States has changed over the past generation, I was referring to the reliance on palliative medicine and the increased likelihood that death will be preceded by a decision to stop curative treatments. However, it is also important to appreciate that the actual diseases that take our lives have changed. As Stephen Kiernan observes in his book *Last Rights*, Americans used to die suddenly and unexpectedly, but now they die slowly. Kiernan makes the point that the principal causes of sudden deaths are heart attacks, strokes, and accidents. In the past three decades, death rates from accidental deaths have dropped by 36 percent; since 1999, the mortality rates for heart attack and stroke have declined by about 30 percent. These improvements are a result of innovative treatments and groundbreaking advances in prevention, such as smoking cessation, implementation of cholesterol-lowering strategies, and use of blood pressure drugs.

Nowadays, Americans die slowly from chronic illnesses such as cancer, diabetes, Parkinson's disease, osteoporosis, congestive heart failure, emphysema, and Alzheimer's disease. Even though we are accustomed to thinking of cancer, for example, as being a rapidly fatal disease, this is no longer the reality. While some types, such as pancreatic carcinoma and certain acute leukemias, still quickly lead to death, the majority of cancers are at least partially responsive to treatment, and patients can survive for years. Alzheimer's disease, because it has no cure, is a slow, progressive killer that's a bit more obvious. In 2007, more than 5 million people had Alzheimer's disease—a 10 percent rise over the American Alzheimer's Association estimate from five years previously—and between 2000 and 2004, the number of deaths from this form of dementia increased by a remarkable 33 percent.

Like most people today, Barbara Dilanian was dying from a number of slow illnesses—renal failure, cancer, and persistent respiratory disease. When people die incrementally, this gives them and their caregivers plenty of opportunities for decisions or interventions. The common pattern of death from slow diseases involves a saw-toothed decline, punctuated by partial recuperations after repeatedly being on the verge of death. While we may not know exactly when individuals are going to die from chronic diseases, the science of prognostication is improving and there is usually ample warning when the overall slope of health begins to fall precipitously. Despite having witnessed numerous people rebound from seemingly terminal episodes, Amy Gleason, Kim Hoy, and other seasoned nurses have long ceased to be astonished when patients finally die.

Amy and Kim are now accustomed to participating in decisions that hasten dying—especially cessations of life support treatments such as dialysis and the withholding of cardiopulmonary resuscitation. This is all part of a paradigmatic shift in the practice of medicine in which mortality is beginning to be more openly accepted. As a direct consequence, people are increasingly confronting the need to make treatment determinations. Put plainly, most people can now have, and choose to have, planned deaths.

There has been a disproportionate amount of media attention accorded to two types of planned deaths—euthanasia and physician-assisted suicide. We do not know how many people die following active euthanasia, but it is certainly a small number. We are able to be much more exact about physician-assisted dying—or at least the officially recorded deaths. Between 1997 and 2008, the Oregon statute enabled a total of 401 individuals to end their lives through physician-assisted dying. There were forty-six deaths in 2006 and forty-nine in 2007. In May 2009, a woman with pancreatic cancer was the first person to die in this way in the state of Washington. However, considerably less publicized are the decisions by *more than 2 million Americans* to die each year following withholding, withdrawal, and limitation of treatment. Citing studies that found decisions accelerating death in 90 to 95 percent of intensive care units, a commentary in the British medical journal *Lancet* has noted, "The good news is that the withdrawal or withholding of life-sustaining treatments is now standard practice."

The pendulum has swung back and forth over the past two hundred years as to how American medicine regards dying. When Philadelphia's Pennsylvania Hospital and Boston's Massachusetts General Hospital were founded in the eighteenth and early nineteenth centuries, they explicitly refused admission to dying patients. The early hospitals were dedicated solely to curing people and did not have the resources to work with the terminally ill, who instead were encouraged to go home and die.

On my office wall is a Currier and Ives print depicting the death of George Washington. He is certainly not in a hospital. He is lying in his bed at home, surrounded by his weeping family, a concerned friend, a solemn attending physician, and two distraught "domestics" (slaves). Great men were ideally supposed to die either in the midst of battle or at home surrounded by heartbroken intimates. However, by the first half of the twentieth century, attitudes changed. Terminally ill men and women were frequently admitted to medical facilities, more funds and scientific research were directed toward medical care, and people developed an expectation that they would receive treatment in hospitals until the end of life. As the population aged, nursing homes became the second most common place to die after hospitals.

In the past, death was rarely seen as being a release; rather, it was considered an implacable enemy to be resisted and overcome. Every explorer who trudged step by step through a hostile wilderness, as well as the heroic figures who fought in the country's wars and military conflicts, all enunciated the same credo—one must continue to fight until the last breath of life. This social context was evident within the discipline of medicine, where throughout much of the last century health care was entirely directed toward thwarting death. Victory was measured by another disease vanquished or by a few percentage points of improvement in a five-year survival rate.

Societal attitudes significantly changed in the last quarter of the twentieth century when a series of landmark judicial rulings signaled a radical shift from the "life at any cost" philosophy. Court decisions involving two young women, Karen Ann Quinlan (1976) and Nancy Cruzan (1988), who each had fallen into persistent vegetative states and died like Terri Schiavo after the removal of life support, served to reify the bioethical principle of patient autonomy and self determination. The most recent study of the Oregon Death with Dignity Act permitting physician-assisted dying has found that 100 percent of the participants cited loss of autonomy as being their personal chief end-of-life concern, followed by decreased ability to participate in activities that make life enjoyable (86 percent) and loss of dignity (86 percent). The Supreme Court has now repeatedly affirmed that not only must patients agree to participate in any treatment, but they can withdraw approval at any time and when they do so treatment must immediately cease. In our modern legal system, treatment without permission is conceptualized as assault and battery by staff upon the helpless body of the patient.

In other words, to a society that tolerates differences and choice of values, the individual is the ultimate decision maker for his or her own body. This has led to the emergence of the far-reaching idea that it is now up to patients and their surrogates—loved ones, families, and medical personnel—to determine if the suffering associated with treatment outweighs its benefits. "Enough is enough" has entered the medical consciousness, and the door has opened to facilitate planned dying. The inspiring seed of modern hospice and palliative care that first appeared in Great Britain has now germinated and rapidly spread throughout the United States.

What are the turning points that have led to the shift in the American zeitgeist and the ready acceptance of palliative medicine? What has led us to a moment in which a dying mother could be told her prognosis and shortly afterward participate in her daughter's wedding staged in the hospital? Although there are many contributing factors, three events warrant emphasis: the groundbreaking efforts of the psychiatrist Elisabeth Kübler-Ross, the substantial breakthroughs in medical technology of the 1960s and 1970s, and, paradoxically, the AIDS epidemic.

It is unfortunate that Kübler-Ross is better known for the appealing but fallacious theory that people go through distinct psychological stages while adjusting to terminal illness. Her theory is still widely accepted by most laymen, but psychologists and psychotherapists have long dismissed the notion that there are distinct and linear phases to coping and grief; in reality, we often simultaneously feel many of the emotions described by Kübler-Ross.

Far more important was her convincing demonstration at the University of Chicago's medical center that gravely ill people are not harmed, but are actually gratified, by being given opportunities to have frank discussions about their fears, aspirations, and terminal care preferences. Breaking with the commonly held belief, Kübler-Ross eloquently explained that the dying do not have to be shielded or protected by a conspiracy of well-meaning lies. She declared that truth does not necessarily destroy hope, and that people respond positively to being allowed the opportunity to take part in decisions. Following her example, other medical practitioners began to listen to their dying patients, and it became apparent that not everyone wishes to survive at any cost. Admittedly, some people want absolutely everything possible to be tried in order to prolong or ensure survival. But there are many others who are adamantly opposed to this strategy.

A second turning point was a series of medical advances that have made organ failure and other severe conditions no longer synonymous with death. It used to be that if your kidneys or lungs failed, you would die. This is no longer necessarily the case. Instead, machinery is available to supplement or substitute for damaged organ systems, and transplantation of heart, lung, liver, bone marrow, and kidneys has become possible. Dialysis for renal failure became not only theoretically feasible but also, with the help of government funding, widely available; dialysis clinics speedily proliferated throughout the country. Intensive care units became omnipresent in hospitals, exemplifying the technological explosion and life-prolonging imperative of medicine.

Patients were surrounded by expensive and sophisticated equipment and dependent on these for the monitoring of vital functions. If one's heart stopped beating, it was possible to plunge a needle directly into it and administer stimulating medications. Alternatively, in a procedure that gave new meaning to the term *aggressive medicine*, one's chest could be "cracked" by a surgeon, who then reached in with his or her hand and manually pumped the organ. On television, as the gentle Drs. Kildare and Welby were joined by the grimmer Dr. Casey and finally the frenetic staff of *E.R.*, the wonders, frustrations, and limitations of medicine became manifest. The public came to gradually appreciate that just because space-age technology could maintain life in the artificial setting of the intensive care unit, dialysis clinic, or surgical ward, all of this might have no bearing whatsoever on whether patients ever returned to anything approximating their formerly satisfying and productive lives.

The onset of the AIDS epidemic also fueled the societal change in our attitude about dying. For the first time since the great influenza outbreak in the early twentieth century, all of medicine's accomplishments were rendered impotent by an apparently irreversible and untreatable terminal disease. Medicine had been in the process of shifting from "do no harm" to "do everything," and this was thrown into serious doubt by AIDS. While fear of contamination and the seeming futility of treatment led some infected individuals to be barred from intensive care units, other patients intentionally chose to avoid the more aggressive medical interventions.

After repeatedly witnessing the debilitation and devastation wrought by both the disease and its treatment upon friends and lovers, a number of people with AIDS chose to actively hasten their own deaths. Suicide emerged from the closet, and it was portrayed as a means to avoid victimization by the infection. Such suicides were often preceded by "living wakes" where friends, family, and loved ones gathered to recall the good times, express whatever remained unsaid, and bid farewell. These were often public affairs—especially in the epidemic's epicenter, San Francisco. The suicides contributed to a growing idea that one need not wholeheartedly agree to battle death using all of medicine's available technological weaponry, and this idea has proceeded to filter its way through American society. It is of course ironic that, given a few more years of medical progress, AIDS eventually became transformed into just one more ordinary chronic disease, and certainly it is no longer an unstoppable plague.

Nowadays in the United States, there are practically no hospitals, nursing homes, medical clinics, or doctor's offices where the option to refrain from aggressive treatments or to cease medical therapies is not discussed, let alone offered. Oncologists may continue to recommend grueling and demanding cancer treatments for patients, but many of these physicians now also entertain the possibility of potentially less *successful*, but less onerous, courses of therapy.

Most Americans have come to agree with Rosalynn Carter's sentiment in the foreword to *The Handbook for Mortals*: "We often put off what is important in life, and it sometimes takes the shadow of death to make us appreciate that love, family, and faith are things that really matter." This was clearly evident with Barbara Dilanian when medical personnel recognized the inevitability of her impending demise and went out of their way to help her and her loved ones make the most of the remaining time. Such deaths are transcendent experiences for participants, and it is essential to appreciate the extent to which they represent a drastic shift from the medical practice of previous generations. They are also a direct challenge to those who believe any intervention that accelerates death is morally wrong.

Opponents of death-hastening practices such as dialysis termination point out that we frequently oversimplify many of these situations. The skeptics are not necessarily wrong with this critique, and both the glory and the complexity of treatment cessation were amply evident in the other case alluded to by Kim—that of Benjamin Babcock.

It's a Wonderful Life

Benjamin Babcock's skin gleamed like the finest mahogany after he received a sponge bath from Olga Vasquez. She carefully applied a lubricating lotion and tried not to inadvertently hurt him. After eleven years of dialysis treatment, his skin had become inordinately dry and gray. Over time, whether you start off as white or black, renal failure slowly changes the color of your skin. The liver metabolizes the amino acids of protein into the waste product urea, which normal kidneys will then excrete into urine. Benjamin Babcock's kidneys could no longer carry out this function. Instead, urea literally sweated out of his skin, evaporated, and took on the appearance of winter frost.

While dialysis periodically flushes out much of the accumulating urea and bodily fluids, it unfortunately does nothing to slow the inexorable progression of aging. The treatment also does little to halt the continuing staccato complications of hypertension, which was originally responsible for causing Ben's kidneys to fail. During each of his subsequent hospitalizations, Olga and the other renal staff members at Baystate Medical Center did their best to help him with any new medical problems. In between crises, they attended to little comforts, such as a warm sponge bath.

My office phone rang and it was a colleague inquiring, "Have you gotten a look at today's *Union News*?"

When I answered no, she said, "Stay put, and I'll be down in a minute."

While handing me the local Springfield newspaper, she exclaimed, "With your interest in dialysis discontinuation, you have got to check this out."

A banner headline read, "Putting Polish on a Gifted Life: Beloved Shoeshine Man in Hospital." Two large photographs accompanied the article. One showed Benjamin Babcock working at his shoeshine stand; in the other, he sat in a Baystate patient room with his aunt, Mrs. Elizabeth M. Ward. The first picture, taken a few years earlier, depicted a handsome black man kneeling in the traditional pose of the profession, vigorously applying polish to a customer's shoes. Ben was clad in a sports jacket with a silk handkerchief poking out of the breast pocket, a matching scarf fashionably draped over his shoulders, and an eye-catching gold necklace hanging around his neck. He was wearing a jaunty cap of the kind that one is accustomed to seeing in illustrations of turn-of-the-century newsboys, and he was smiling benignly while he worked. The photo drew attention to Ben's large and gnarled hands. You could sense his power, mastery, and satisfaction.

In the second picture, Ben looked much older and frailer. He had visibly shrunk. The sports coat had been replaced by a woolen robe, a blanket was spread on his lap, and he wore hospital pajamas. The same gold necklace hung from his neck. Ben's aunt protectively clasped his hand, and they looked at each other with obvious affection, but also considerable uncertainty.

The newspaper article recounted that Ben had decided to terminate his dialysis treatment, and streams of people had been coming to his bedside to say goodbye. At least a hundred people appeared on a single day, necessitating that he be moved to a larger room. The hospital estimated that over the course of a couple of days Ben had more than three hundred visitors.

The *Union News* article began, "He isn't a senator or a millionaire or a movie star. He is a shoeshine man." One of his longtime friends was quoted as saying, "My feeling is if I touch one-tenth of the people that Ben has touched, then I know that my life on earth was worthwhile." Another friend remarked, "Ben has an incredible gift. He has enriched a lot of people's hearts with his heart."

Eleven years previously Ben's kidneys had ceased to function, forcing him to leave his position as head steward at the Valley Athletic Club of the YMCA and to instead earn his living shining shoes. Until a few months before the final hospitalization, he could mostly be found at his stand in Tower Square near the Springfield courthouse, taking his familiar place amidst colorful bottles of Kiwi shoe polish.

Ben was quoted as saying, "The [dialysis] treatment was fine. But other complaints set in that give me problems. . . . In the process, I became sick and tired of going through the traumas. I have the option to continue with the treatment or terminating it, and that's what I chose to do."

This assessment was echoed by a neighbor, who for the past three years had been driving him back and forth to dialysis. She later told me Ben always brought along a small radio to listen to his beloved jazz, and staff regularly positioned his dialysis couch next to young patients. The music would form a backdrop as the men and women—often in their early twenties—told him their concerns, and Ben would encourage them to cooperate and endure the treatment. Now, between his illness, complications, and exhaustion, the friend observed, "He had broken down."

The newspaper article described Ben patting his aunt's hand and saying, "Goodbye, Betty." Calling across the room to his daughter, Sandra E. Babcock, he intoned, "Sandra sweetie, this is so long." And then Ben began singing the Benny Goodman song "Goodbye."

One of the hospital visitors was U.S. district court judge Michael Ponsor. Over years of regularly having his shoes shined by Ben, he had become a friend. The judge was quoted as saying, "The prospect of not having him with us really makes me sad. . . . There are certain people in the world who spread light wherever they go, and he is one of them. . . . Whenever I ran into him, saw him, or talked to him, I came away saying, 'I'd really like to be like that.'"

Ben's daughter admitted to being initially uneasy with the decision to stop treatment. But she went on to say, "Now I see all these beautiful people come to visit and their lives were touched. I keep hearing in my head, 'The miracle is accomplished.'"

When I walked down to see what was occurring, I observed Benjamin Babcock's primary nephrologist, Dr. David Poppel, talking with Kim Hoy at the renal unit's nursing station. Amy Gleason had just crossed the hall and was bringing a new transplant patient some medications to prevent kidney rejection. Kim and Amy had both ministered to Ben during previous Baystate admissions, and they were not surprised that he had decided to stop dialysis. However, they were amazed at the reception he was receiving from the community. The two nurses were quite aware of the unusual amount of excitement on the unit, and they were pleased that his decision was being not only respected but celebrated.

Poppel was placing orders in the computer when I caught his attention. Moving over to the "bubble room"—a quiet spot down the hall that earned its nickname from the bank of windows that allowed any passerby to look in on the interior—he and I sat down together.

Poppel is universally acknowledged to be a likeable guy. I have been collaborating with him and his colleague, Dr. Mike Germain, on a series of palliative care studies involving dialysis and kidney transplant patients. David is a handsome, open-faced man with lustrous brown hair who wears thin round spectacles and frequently clomps around in cowboy boots to signal his fondness for horses, the open plains, and the rustic beauty of Montana. I often smile when I see David, because he reminds me of Gene Wilder's character in the movie *Blazing Saddles*.

On that day he was looking thoroughly professional, though, and I inquired as to how Ben had reached the decision to discontinue dialysis. David replied, "Last weekend I was in the hospital and came around to see him. I was told that he wanted to speak to me, because he intended to terminate his dialysis treatments. According to the nurse, he had already spoken to his family and prepared everyone for the decision.

"Well, I went into his room, and he was very at ease, alert, and observant. Because of certain things occurring in his medical condition, he recognized that he was entering a new phase. In the past, he had his ups and downs—as many dialysis patients do, especially as they get older. There were times in the hospital when he was not doing well physically or emotionally, but he was always able to sense he would be able to rebound—either on his own or with the help of his caretakers—and come back to a baseline that was satisfying. However, I believe he finally realized that there was something beginning to happen with respect to the circulation in his feet, some infections that were developing, and pain. . . . He was entering into a new phase in which a series of difficulties were likely to be encountered on a regular basis. I think he had seen the same thing happen to other people—patients who wanted to 'tough it out' or who were just not aware it was the beginning of the end, and they could and should avoid the hardships that would follow."

As Poppel recounted it, Ben had told him, "I am starting

something new. Why should I suffer? Why should my family suffer? Why shouldn't I just come to grips with it now?"

It was a bold statement, but one that Poppel was not shocked by.

"Quite courageously, Ben had come to think about it in new terms, and his solution was to say, 'Dialysis has been useful to me. It has helped me stay alive when otherwise I would not have lived and enjoyed my family, music, and work.' But now he wanted to stay comfortable. His close friends and family concurred with this decision. What he wanted from me was information about dialysis discontinuation. He wanted to know, how do you die? How long does it take?"

David glanced out the windows of the bubble room and he and I took in the familiar sight of staff, patients, and visitors walking by. Olga was in the corridor tenderly helping a patient to take her first uncomfortable steps after a surgical procedure. David looked back at me and said, "Ben is doing very well. He is in a comfortable, private room and is mentally sharp. He is having many visitors and is clearly enjoying himself. He is right on top of this. Between friends, family, and the media attention, the experience is exceeding his expectations. He is a jazz aficionado, and his guests are talking about music. You can hear a boom box softly playing selections from Miles Davis and John Coltrane."

I inquired what David's thoughts were about the decision to terminate treatment. He answered, "I have known Ben for ten of the eleven years that he has had kidney problems. He is insightful; his decision is both courageous and correct. The way he arrived at it satisfies him and the people who care for him. I am content to pop in every day—whether in my official capacity or simply as a friend—in order to say goodbye."

I am a complete sap about the perennial Christmas movie *It's a Wonderful Life*. Each year, to the great scorn of my family, I watch it on television, revisit its characters, and enjoy the sensation of tears dripping down my cheeks. I am an object of special derision to my wife as I loudly sob and savor the film's emotional high and low points. At the beginning of every December, my sons promise to watch this cinematic treat with me, yet somehow each year I end up sitting alone in the living room.

For those unlucky few souls (including my children) who have never seen Frank Capra's masterpiece, the relevant part is that George Bailey, the movie's protagonist (played by Jimmy Stewart), is overwhelmed by an unfortunate confluence of events that will result in bankruptcy, and he staggers toward a snow-covered bridge with the intention of ending his life. George is saved by a hapless guardian angel, who gives him an opportunity to revisit the past and see the many people who have been positively influenced by his existence. The movie concludes with our hero deciding to continue living and the entire town of Bedford Falls arriving at his home to celebrate. One and all declare gratitude to George for contributing to their lives, his financial problems are resolved through communal generosity, and his heroic brother returns from World War II in time to toast him for being "the richest man in town!"

A couple of days after my initial visit, I returned to Benjamin Babcock's hospital room and discovered that he, too, had chosen life. "I'm back on dialysis," he announced to me, explaining that he was so touched by the outpouring of affection from the community of Springfield that it no longer made sense to terminate the dialysis. He clearly felt like the richest man in town.

So I was entirely unprepared a few minutes later when I saw Ben's aunt and daughter gently but firmly confront him about his change of heart. Ben was peacefully propped up in his hospital bed while his family and I sat around the perimeter of the room. Reaching out to take his hand, Mrs. Ward remarked, "Ben, you know that I originally did not agree with your decision to stop dialysis. Since then I have had a chance to see for myself just how sickly you have become, and I now think that it was the correct decision. You are excited, because you have come to realize over these last days just how much you mean to all those people. But let me point something out to you. We are your family, and we love you, and we are here with you. Those other people are not here now. They may love you, too, but each day there have been fewer of them coming to the hospital. We are certainly all right with whatever you choose, but I think that nothing has really changed with your health since you made your original decision. It's hard for me to say this, but I think you were correct in stopping the dialysis."

After undergoing a couple more dialysis sessions, Benjamin Babcock quietly decided to terminate the treatment again. A few days later, his obituary was published. It bore the simple title "City Shoeshine Man." In it were quotes from family, friends, Judge Ponsor, and a U.S. congressman. The obituary accurately acknowledged that Ben stopped dialysis twice before finally dying. His family suggested that memorial contributions be made to the National Kidney Foundation.

I discussed Ben with Olga Vasquez, the nursing assistant, as well as with Amy Gleason and Kim Hoy, the renal nurses. Olga had a complex reaction to his death. She clearly did not approve of dialysis discontinuation, but she felt great affection for this patient. She told me, "I so admire Ben. He reminds me a little of the book *Tuesdays with Morrie*. He wanted everybody there. He didn't know how to separate. . . . He was a wonderful person."

Amy and Kim were in agreement that Mr. Babcock had the right to change his mind about such a major decision. But they were equally clear that in their opinion ending the treatment was the correct choice. They do not believe in prolonging torment.

Fast-forward ten years. Serendipitously, I found myself interviewing one of the many individuals who had come to see Benjamin Babcock in the hospital. At the medical center I am responsible for evaluating all potential transplant organ donors, and this woman was highly unusual because she wanted to provide her kidney to absolutely anyone who was in renal failure. This anonymous, altruistic donor had originally been a member of Ben's Alcoholics Anonymous group. When he was admitted to Baystate, she came to visit him accompanied by twenty other people from AA, who stood in a circle in the hospital room, joined hands, and prayed. Her resolve to help someone with chronic kidney disease was rooted in that visit. She recalls thinking, "If only someone had donated a kidney to Ben, maybe he would have lived a longer life. I wish things could have been different for him. It was a shame he had to die. It was a shame he came to the decision that he no longer wished to live."

I was touched by her sentiment and chose not to challenge her impression that Ben had been a potential transplant candidate. In fact, his hypertension and other health problems had wrought sufficient damage that it would no longer have been feasible for him to undergo major surgery. He was facing the need for another amputation when he made his decision, and there was some question as to whether he could have survived that surgery, and if so, to what extent he would have recuperated.

So, a decade after Ben's death, my feelings remain bittersweet. I am appreciative that the American medical system values autonomy and will not force people to endure unending treatment. I am pleased that the care we provide is flexible and responsive to patient preferences. I am also extremely moved that an unexpected part of Ben's legacy might possibly allow a complete stranger the opportunity to have a kidney transplant. On the other hand, I am perturbed that Benjamin Babcock was sufficiently ambivalent about his decision to terminate treatment to have undone it—and then undone it again. This reversal highlights the complicated nature of a patient being able to choose the timing of his or her death. When it comes to medical determinations that accelerate death, the preference is to see consistency on the part of the patient and family—but that doesn't always happen. I would have wished for Ben's last days to have been more straightforward. However, I also know that for many years his illness had swung wildly between bouts of desperate disease and hard-fought rehabilitation, and that his overall life story similarly zigzagged back and forth.

8

Discontinuing Dialysis

The decision to cease or withhold a life support treatment such as dialysis is both momentous and intricate. Most people do not appreciate the frequency of these mainly private decisions, and we rarely hear about them aside from the well-publicized cases of a few celebrities, such as that of the humorist and Pulitzer Prize–winning author Art Buchwald. Yet today, 26 million adult Americans have chronic kidney disease, more than 100,000 have had transplants, and the lives of some 350,000 Americans are being prolonged by renal replacement therapy. All of these individuals are intensely aware that if they stop taking medications or undergoing dialysis, they will perish; they live every day with this reality. Not all of them are informed that discontinuing treatment is an option, and many would never consider this course of action under any circumstances.

Kim was absolutely correct in her impression that dialysis is an arduous therapy. While prolonging life, the treatment does nothing to halt the progression of disorders such as diabetes or hypertension that cause kidney failure in the first place. Despite the best of intentions, great expense ($32 billion annually), and often Herculean efforts, each year more than one out of every five patients maintained with dialysis in the United States will die. This death rate is higher than that for AIDS, colorectal, breast, and ovarian cancers. It is always a leap to go from statistics to specific cases, but on average, if you are receiving dialysis, you can expect to live between one-quarter and one-fifth as long as your counterparts in the general population when matched for age, race, and gender. And so, given the often grim prognoses and suffering these patients face, more than 20,000 annually choose to terminate dialysis and accept death. In about half of the cases, individuals no longer have the capacity to meaningfully make determinations, and family and staff members arrive at the difficult decision on their behalf. In New England, where I practice, more than four in ten deaths of dialysis patients now follow discontinuation of treatment.

For a number of years I have been trying to make sense of why people decide to stop dialysis, and I now believe the decision begins with awareness on the part of a patient and his or her caregivers that the process of dying is well under way. This is coupled with an unwillingness to relinquish the attributes and pursuits that an individual considers integral to staying alive.

David Poppel's assessment was that Benjamin Babcock knew something had changed in his body—and this makes sense to me. While it is true that anyone with renal failure who is maintained with dialysis will die if the treatment is stopped, when you speak with otherwise vigorous individuals (especially young ones), it is evident that they are not and do not necessarily view themselves as being "terminal." There comes a point, however, when some people sense their bodies are giving out—it becomes apparent to them they are dying and it may no longer seem logical to protract matters.

In my experience, decisions by patients with renal failure to cease life support treatment are frequently associated with a particular personality. Benjamin Babcock is a good example of that personality—strong-willed, fiercely independent, and charismatic. Art Buchwald had a similar personality. Physical illness eventually brings these vibrant and attractive individuals to unrecognizable places where everything becomes indelibly transformed by suffering. However, when they arrive at the decision to end their lives they do not appear to be driven by despair or clinical depression. Their determination mainly reflects the practical and active ways they customarily manage situations. I think of them as being people who recognize that they have passed their train station and want to disembark as soon as possible.

Woody Allen has quipped, "I don't want to achieve immortality through my work. I want to achieve it through not dying." While I appreciate that most people completely concur with the comedian's sentiment, I am not one of them. Instead, like most palliative medicine physicians who work with dying people, I have long ago stopped viewing death as an evil to be avoided at any cost. I share the opinion of a colleague who wrote in an editorial addressing why doctors should communicate poor prognoses to patients, "Sometimes living life to its fullest requires knowledge of its finitude."

The kidney is a master organ when it comes to regulating fluid, controlling blood pressure, and excreting various toxic substances. My longtime research collaborator, Mike Germain, is a kidney specialist—a nephrologist—and like most nephrologists, he is obsessed with the various chemical and blood indices of his patients. Since kidney doctors care for a population that is increasingly elderly and disproportionately diabetic and hypertensive, they need to master many aspects of geriatrics and general medicine. They are also almost unique among physicians in that they see most of their patients three times per week for multiple years. Consequently, patients often look to them to be their primary care doctors.

Although renal replacement therapies such as hemodialysis, peritoneal dialysis, and kidney transplantation can extend people's lives for decades, patients with renal failure still have a markedly diminished life expectancy as compared to the general public. Accordingly, nephrologists are repeatedly faced with death and dying issues, and decisions like those that Benjamin Babcock and Barbara Dilanian made will continue to occur with increasing frequency.

Mike is fortunate in having a precious ability to simultaneously enjoy being an eggheaded scientist while also engaging in the more emotional aspects of medicine. During one of our meetings I asked him if we could discuss the topic of dialysis discontinuation, and Mike promptly suggested we role-play the conversation that might take place between himself and a patient. Mike has no idea how much I dislike role-playing. Nevertheless, I said, "All right. I'm a seventy-year-old patient of yours who has complicated diabetes. It is now four years into my dialysis treatment, and I am seriously considering stopping. I would like you to tell me, in some detail, what would be involved."

Mike began, "We've talked about this a little bit in the past, but I know a lot of times you forget things." He smiled playfully at me. "Also, I am glad that your family is meeting with us today, because I am sure your wife and children have questions and would want to know what is involved in this process.

"If a patient stops dialysis, they usually survive somewhere between seven and ten days. Generally we find that people have a peaceful death. They can be made quite comfortable. I appreciate that right now, despite my best efforts to provide you with pain medicines, your legs have been bothering you from the vascular disease. Once you stop dialysis, we can really push up the pain medicine and try even harder to relieve the discomfort. Stopping the dialysis itself should cause no additional suffering."

"And where would I be while this went on?" I asked.

"I'm glad you brought that up," he quickly responded, "because it is an important thing for you and your family to think about. Right now you are in the hospital. Some people choose to stay here and are comfortable with that decision. Other people prefer to go home. This last choice requires your family being able to manage matters. In either case, we can get hospice to lend a hand. Here in western Massachusetts we are fortunate to have a hospice that can assist whether you are in the hospital or at home, and there is also a hospice residential program where you could be transferred. Finally, following some of your previous admissions you have been sent to one of our local nursing and rehabilitation centers, and it would again be possible for care to be arranged there. Have you or your family thought or talked at all about the different options?"

I chose not to answer, but instead asked, "What is it actually like at the end? My family wants to know what they can expect."

He answered, "We will go into this in more detail at a later point. For now, you and your family should know that during the first few days you are likely to remain alert and hopefully be more comfortable. You will be able to continue to meaningfully interact with them and with your visitors. Gradually you will get sleepier and probably slip into a coma. When you are no longer awake, there may be some gurgling, twitching, or gasping. We will reassure your family that these are normal things and they do not mean that you are experiencing any discomfort. Sometimes we can give medicines to alleviate those signs, but we do not believe they reflect any actual suffering. We have made a commitment to be available and to take care of you. Whether or not hospice is also involved, we will be there for you and your family."

Despite Mike's assurances, the reality is that physicians do not know if the terminal signs and manifestations are uncomfortable. How could we? His commitment to be present is likewise more of an ideal than a guarantee. He and I have both seen situations in which family members began a deathbed vigil only to discover the patient had expired when they briefly went to the cafeteria to grab a bite or slipped out to go to the bathroom. Likewise, for many unforeseen reasons, medical care does not always remain consistent during the final days or weeks.

"I understand that you were recently at a patient's home when he died," I said, changing the topic a bit. "Would you tell me more about it?"

Mike looked a little uncomfortable but, staying in his role, he said, "I actually enjoy visiting people at their homes. I would be willing to do that with you, too. I don't do it with everyone who stops dialysis, because there are situations where the person does not necessarily want the doctor to come or he may not be conscious enough to appreciate a visit.

"The person you are alluding to had been a patient of mine for quite a few years. When I visited him, he had a hospital bed set right in front of a picture window in his house in Wilbraham. He could look outside to the garden and seemed very happy and peaceful. His family had gathered together from all over the country. This included a ninety-two-year-old sister who arrived to say her last goodbye. I asked him and his family whether there was anything I could do for them. Was he comfortable? Was he having any symptoms? After that, we talked about personal things. You know, about his life—such as what it was like when he was younger. His family brought out some old pictures, and we were chatting about them when he died. We talked a little bit about what he did in World War II and how he got married. He was drifting in and out of sleep, and he fell asleep and did not wake up. There was no perceptible difference between when he was alive and when he was dead. It was that peaceful. His family was watching us, and when I told them that he had died, they said, 'You have got to be kidding!'"

I inquired, "You were holding his hand at the time?"

Slipping entirely out of his role-playing mode, Mike quietly said, "Yes. I think it's nice to have . . . you know . . . physical contact. They like . . . people like to be touched, especially in their last few days. You know . . . I think that is something that I find valuable, and so do my patients. However, in this case I was as surprised as the family. One moment he was alive and I was holding his hand . . . he seemed a bit drowsy . . . and then in the next moment he was quietly gone."

Bobby Schindler and other like-minded people will read about Benjamin Babcock's story and speculate that he wanted to kill himself—that he was depressed and decided to commit suicide. They will place emphasis on his ambivalence and the manner in which he made and unmade his decision to terminate treatment. While I do not agree with this position, I think that it merits consideration.

In 1971, a pioneer in psychiatry, Dr. Harry Abrams, wrote about his experience participating in the care of the first generation of patients to receive dialysis. His belief at the time was that patient deaths following treatment noncompliance or dialysis discontinuation were suicide equivalents, and he published an article warning that the resulting suicide rate was four hundred times that of the general population. However, by 1977 Abrams had begun to question his own logic, and he eventually joined other psychiatrists in expressing the belief that cessation of life support and suicide were two separate phenomena. More recently, Drs. Norman Levy and Adam Mirot have stated that rational motives for a patient to refuse continuation of dialysis are legion. Especially if they are not transplant candidates, dialysis patients suffer significant discomfort, inconvenience, and progressive functional disability. The return on all this is a possible prolonging of life, albeit on a limited scale. It is understandable that the risks and injuries attendant to long-term dialysis may eventually outweigh the perceived benefits, and these two psychiatrists maintain that, under such circumstances, withdrawal from dialysis is appropriate and permits the facilitation of a good death—characterized by comfort, dignity, and brevity.

One of my good friends, Dr. J. Michael Bostwick, is a suicidologist from the Mayo Clinic. Michael emphasizes that in the context of terminal illness—in which life is artificially prolonged through the use of technologically advanced treatments, such as dialysis—it is instructive to examine the social setting and reactions to withdrawal. In traditional suicides, surviving acquaintances and family are invariably appalled and devastated, whereas in cases like that of Benjamin Babcock or Barbara Dilanian, the bereaved loved ones are saddened but accepting and sometimes even proud of the deaths. Michael and I are in agreement that these constitute two distinct types of deaths.

The topic of suicide is emotionally charged and historically laden with legal and religious baggage. The historical background for suicide has best been described by A. Alvarez in *The Savage God*. He pointed out that for many generations people who "successfully committed suicide" (always an odd choice of words) were believed to have made a pact with the devil. In England, their bodies were denied normal burial and they were instead publicly interred at the crossroads—usually with a stake driven through the heart to deter their ghosts from walking around or returning to haunt homes. Crossroads burial was only abolished by an Act of Parliament in 1823. In many countries, including the United States, people who attempted to kill themselves were considered to have perpetrated a criminal act. Although no longer true in America, suicide is still illegal, for example, in India. (However, the Law Commission in New Delhi is actively trying to change this antiquated penal code.)

The field of law has provided the best conceptual framework by which to understand the relationship between withdrawal of life support and other acts that accelerate dying. American jurisprudence has explicitly determined (in a New Jersey state supreme court ruling involving the case of Claire C. Conroy) that "declining treatment may *not* be properly viewed as an attempt to commit suicide, as the refusal merely allows the disease to take its natural course and the subsequent death would be the result primarily of the underlying disease and not the result of a self-inflicted injury."

Although most of us think of the law as being a rather blunt instrument, court cases often must address and assess subtle differences. For example, jurisprudence clearly recognizes that all killing is not the same and does not warrant identical punishment. *Homicide* simply connotes the death of an individual at the hands of another. Whether or not a homicide is punishable—as well as the extent of the punishment—depends mainly on the mental culpability of the person causing the death. Intent is a key factor in these determinations. For example, it is obvious that a soldier who kills on the battlefield or a citizen who kills in self-defense are engaging in behaviors that differ from that of a bank robber who shoots a teller. For this reason our legal system wisely differentiates between types of homicide, including first-degree murder, second-degree murder, justifiable homicide, manslaughter, and so on.

Given that the law recognizes a spectrum of acts of killing—one that is shaped by motivation and intent—we might do well to also broaden our concept of suicide to include a spectrum of life-ending acts that are similarly shaped. More than one hundred years ago, the sociologist Emile Durkheim made the point that some suicides are not only acceptable to society but highly laudable. It remains for us to devise a suitable vocabulary to label these new categories. New terms are needed that do not contain the pejorative word *suicide*.

Palliative care clinicians are attracted to their work by a compulsion to ameliorate distress and a need to help people have good deaths in a morally acceptable fashion. They highly value their own communication skills, and they intentionally share the burden and responsibility of caring for the terminally ill by espousing an inter-disciplinary team approach. Clinicians earnestly want to listen to what patients and families desire as death approaches, and autonomy is held to be the preeminent bioethical principle. Much of the satis-faction in the practice of this kind of medicine comes from trying to elicit individual patient preferences and attempting to achieve these despite the worst manifestations of disease and organ failure. Treat-ment is not considered to cease with the patient's death, but rather to continue through the provision of bereavement support for the loved ones. It is anathema when these sensitive practitioners are la-beled as being agents of death, although my research has found that palliative care clinicians are quite commonly the butt of humorous jibes of exactly this sort (my own family has been known to call me Dr. Death). Physicians and nurses quietly debate among them-selves whether their actions hasten dying, but most staff members earnestly maintain that they do not. Absolutely none of these heal-ers think of themselves as being killers.

There are several reasons that palliative medicine practitioners believe their actions do not accelerate dying, and chief among these is the principle of double effect. It is practically impossible to talk to palliative care clinicians who are considering either administration of high-dose opioids or withholding and/or stopping life support treatments without them citing this misleadingly simple principle, which was first articulated by St. Thomas Aquinas in the thirteenth century and continues to be regularly offered as justification for modern medical practices. According to Aquinas, some actions have both positive and negative effects; they are acceptable if the good consequences and not the bad are intended. The principle goes on to say that the bad effect must be tolerable, foreseeable, and unavoid-able, and that the good should outweigh the bad.

Aquinas' principle of double effect is accepted in law as well as in medicine. Life itself is filled with instances of double effect, such as the collateral damage seen in battle, or the workers in a struggling business who lose employment when their company is bought by a larger, more efficient firm. It is an everyday occurrence in medicine when complications ensue during surgery or people suffer side effects from medications. I give neuroleptic medicines to my delirious and psychotic patients to help them, all the while knowing that a certain unlucky percentage will then have (treatable) muscle spasms or tremors, a few will develop a persistent neurological disorder called tardive dyskinesia, and an even smaller number may perish from a dire side effect called neuroleptic malignancy syndrome.

Although I salve my conscience by occasionally relying on this theory, which I use in explanations to patients, families, students, and fellow staff members, it still does not leave me entirely satisfied. My colleague Dr. Walter Robinson from Dalhousie University in Halifax, Nova Scotia, has similarly concluded that there are many clinical situations that do not exactly meet its criteria. He explains, "[The] Doctrine of Double Effect is philosophically weak, it is overused, and it may not express clinical realities. But, it is an easy thing to teach, and it may get people to treat pain when they otherwise would not." In other words, it is useful in its simplicity, but not in its accuracy.

The single most archaic and clunky aspect of Aquinas' principle has to do with its oversimplification of intentions. Even the least insightful and psychologically astute individuals in our society can appreciate that people rarely have clear, straightforward, singular intentions underlying their actions; people are too complicated to do things for only one reason. A century ago, Freud made the point that a particular character in a dream may appear to physically resemble your mother but simultaneously represent several people of different genders and differing points of view. Similarly, in our waking life we also inevitably have multiple intentions, some of which are conscious, and others of which are just as pertinent and may be even more significant but are entirely unconscious.

When it comes to the treatment of dying patients, doctors and nurses bring to the bedside complex intentions and beliefs that often have been shaped by the deaths of their own loved ones. Practitioners do not want people to suffer unduly, and they want to respect patient wishes or those of their surrogates. But they are influenced by multiple competing desires, including a strong preference to not extend agonizing situations for patients, families, and themselves. Furthermore, while doctors in the United States are not obligated to provide the least expensive and most fiscally responsible treatment, they are not oblivious to the growing imperative to appropriately allocate resources. Physicians cannot help being affected by crowded hospital corridors and the lack of available beds in intensive care units. If this does not impact them directly, it has an effect on the medical centers or clinics that provide their employment. As mentioned earlier, Dr. Steven Miles suggests that 85 percent of the 2.4 million deaths occurring each year in the United States medical system are preceded by decisions to limit life-sustaining treatments. Miles does not see this being significantly reversed in the future, because he is convinced that America simply cannot afford to offer more expensive care. Medical treatment at the end of life already consumes 27 percent of the Medicare budget and 10 to 12 percent of the estimated $2.25 trillion total national spending on health care annually. Money has been and always will be one of several motivating factors in medical decisions. I hope, though, that it will never become a decisive motive underlying most treatment determinations.

9

Understanding Olga

Palliative practitioners strive to help people die like Barbara and Ben. But these ideal scenarios are not feasible in every situation. There are many instances where a force, be it medical, familial, or legal, complicates matters. In the case of Rosemarie Doherty that force was Olga Vasquez.

Olga was a witness to circumstances that could not be ignored, and she told me, "It was something you don't get over. They took morphine from the Pyxis that belonged to another patient . . . they said they wasted some, but no one saw this."

A Pyxis machine is a locked computer-driven device that contains medicine—typically, controlled substances. It is usually located on the ward; nurses punch in a code and type in the patient's name. They manually instruct it as to what medicine is wanted and a drawer opens. Staff count what is in there, take out the medicine, and determine how much is left when theirs has been removed. They then close the drawer, and the machine keeps track of everything.

Olga believed that narcotics were diverted, and she was aware that the nurses signed out medications for each other and did not witness drugs being wasted. She could not understand why Rosie was given morphine when "she wasn't screaming in pain."

Olga explained, "It was horrific—not to be believed! Ever! Ever! Ever! That lady wanted to wait for her kids to come and see her. She never had that chance."

At one point during a telephone interview, Olga asked, "Did they [the nurses] think I was a nobody? I was a TA. . . . Did they think I was taking it the wrong way?"

Olga's comments poignantly underscore that although hospitals rely on a team approach, there is a still a strict hierarchy remaining in effect. Staff often choose to politely ignore the intersection of social class, education, ethnicity, religion, gender, and other factors that may play a role in determining one's place in the organization. The chances are that Olga's grievance about being ignored or about her opinion being dismissed is probably not just in her head.

Though she may not realize it, in my opinion, Olga has a number of philosophical allies in the hospital, including Dr. Richard Wait, chairman of the Department of Surgery. Richard specializes in performing surgery on patients with pancreatic carcinoma—a disease legendary for its poor survival rate, and which requires an arduous procedure that routinely takes five or six hours to perform.

Richard believes doctors have different thresholds for deciding a patient's condition is futile, and he has jousted with more than one colleague when he thought they were prematurely abandoning life-prolonging treatments. One such disagreement took place with an intensive care unit physician who decided not to follow Richard's medication recommendations because he determined the patient was dying. The physician went so far as to convene a family meeting to discuss treatment withdrawal without including the surgeon. According to Richard's version, he was forced to confront the ICU staff member and thereby save the patient's life.

With a trace of pride, he calls his aggressive type of pancreatic cancer surgery "assault and battery." He told me, "Surgeons know what they have done to you, and they know what outcomes to expect." According to Richard, "I don't think physicians who are not well trained in taking care of these kinds of patients are in the best position to make decisions about the end of life. I think such a decision should be made with the person who has been there before the operation, during the operation, and after the operation." In other words, doctors should not be in such a hurry to declare situations hopeless, and when surgery has been a major part of treatment, surgeons should preferably be the ones to determine when to call it quits.

This leads us to consider the troubling mathematics of serious disease. Patients with untreated pancreatic cancer live for an average of eleven months; major surgery can increase this figure to twenty-two months. Richard pointed out that during the additional time, people can wrap up their affairs, reconcile with loved ones, and realize personal aspirations. On the surface it sounds eminently sensible, but I am still skeptical.

During our interview, although I do not have cancer and we were talking physician to physician, I nevertheless refrained from asking a number of pointed questions, such as whether there is any way to predict or measure the toll one is likely to pay undergoing this type of major surgery. In our conversation the astonishingly high surgical complication rate of 40 percent came up, and it would seem crucial to know the likely impact of any of these new problems on patients' remaining time. How many days are going to be spent in the hospital on a ventilator? Will patients be able to return home, will they remain in the hospital, or will they be forced to reside in a nursing facility? Is there any way to predict symptoms and quality of life following the surgery compared to what is likely to happen to people who refrain from the operation and instead receive comprehensive palliative care? I am not sure about the extent to which Richard and his colleagues discuss such matters or whether there are substantive data to answer these questions. I am sure the power differential that I experienced in his office, and which muffled my impulse to ask the questions, is felt even more intensely by patients and families in their meetings with surgeons.

As Richard discussed these acts of prolonging life through surgery, he did not mention the case of Dr. Michael E. DeBakey—perhaps the world's most famous heart surgeon—but DeBakey's story is worth describing, as it spotlights the risks and pressures around withdrawal of treatment decisions. In DeBakey's case, his wife and his fellow surgeons overrode all objections (including DeBakey's own) and in so doing were responsible for him undergoing a major and ultimately lifesaving operation. In 2006, the ninety-seven-year-old DeBakey diagnosed himself as having a dissecting aortic aneurysm (a tear involving the main artery of the body) and then chose *not* to undergo the surgical procedure he had devised and which had been previously used on more than ten thousand patients. For nearly a month DeBakey refused to be hospitalized. Admitted under emergency circumstances, he rapidly became unresponsive and incapable of any further participation in treatment decisions. The hospital's anesthesiologists cited a written directive signed by the celebrity patient declaring he did not want to have the surgery, and they refused to assist in any proposed operation.

However, a surgical team that included his partner of forty years insisted that the great man's life could be saved. This view was championed by DeBakey's wife, who barged into a meeting of the ethics committee to demand that the staff accede to her wishes and immediately begin the operation. Before being discharged nearly eight months later, DeBakey had spent six weeks on a ventilator, required a stomach tube for artificial feeding, and had been dependent on short-term dialysis for acute renal failure.

According to Lawrence K. Altman's story in the *New York Times*, DeBakey later remarked that despite his clearly articulated preference to the contrary, his surgical colleagues were correct in performing the operation. It is difficult to argue with his logic—"If they hadn't done it, I'd be dead"—yet there could hardly be a clearer example of old-fashioned medical paternalism, nor could one find a more blatant clinical decision that was based on ignoring patient preferences and autonomy.

DeBakey's statement reflects a standard in which the ends are used to justify the means. And not only did the surgery prolong the patient's life, but after recuperating from his medical ordeal he was able to accomplish at least one more important professional goal. In October 2007, Dr. Denton Cooley presented him with a lifetime achievement award at a meeting of Cooley's Cardiovascular Surgical Society. The true significance of this event was that the two brilliant surgeons had been engaged in a rancorous public feud—they had barely spoken to each other in nearly fifty years—and their rapprochement delighted heart specialists at the meeting and around the country. Michael DeBakey was ninety-nine years old when he died in July 2008.

Joyce Smith, a psychiatric social worker, is another person at Baystate who shares many of Olga's concerns. Joyce believes that we are far too quick even to *discuss* cessation of life support treatments, let alone actually stop them.

"I think the medical profession is almost blasé about the whole issue of decisions about end-of-life care," she told me. "I don't think that the public is quite there with us. For example, I have seen several people come to the emergency room to be admitted, and a physician pops in and says, 'If you stop breathing, do you want us to do everything to bring you back to life?' Well, if my mother was that patient, she would be horrified. Despite the publicity about health care proxies, advanced directives, and other things, I do not think that the general public is expecting to be asked these questions—especially when they are in the hallway on a stretcher.

"There was a patient I knew quite well who had been sick for a long time and had many complications. The patient had fought through all of them, and as far as I know she intended to continue fighting anything that came her way. Well, she took a turn for the worse and ended up on emergency dialysis. It looked like she was facing long-term dialysis. The patient was miserable at this point and clearly felt very discouraged. I left on a Friday, and when I came back on Monday the woman had died. So I asked about what transpired. I learned that she was approached by our well-meaning nursing staff, who said, 'You do not have to suffer like this. You don't really have to undergo dialysis if that is not what you want.' I understand that her family was initially opposed to stopping, but over the course of the weekend and after several conversations, everybody arrived at the decision to end the dialysis.

"Personally, as someone who had known this patient over several years, it felt like the whole process went way too fast. How will I ever know? Well, I won't. What I worry about is whether in our wish to be helpful to patients and to curtail suffering, we may be overly aggressive about making treatment termination recommendations. We may be stampeding patients and their families."

Joyce described another troubling case of a catastrophically ill patient that the hospital staff believed would never be successfully weaned if intubated and placed on a ventilator. After the physicians attempted to convince her that she should not have the treatment and instead ought to agree to do-not-resuscitate status, the patient became upset and a psychosocial consult was requested. Joyce's assessment was that this was the kind of woman—much like her own mother—who regardless of the medical facts would never be ready to make a DNR decision. It made more sense to Joyce to place the woman on the ventilator. If she could not easily get off, then the medical professionals would need to assume responsibility for eventually stopping treatment.

Joyce's comments about advance directives echo developments that are taking place to improve communication of patient preferences. One entails a greater reliance on patient surrogates or proxies to speak up when people have lost the capacity to actively participate in their care. Unfortunately, this approach has a number of limitations, including that it will never satisfy disputes among family members or friends as to who should be the arbiter and spokesperson. Another innovation is the Physician Orders for Life-Sustaining Treatment (POLST) Paradigm Program, which was first instituted in Oregon in order to overcome the limitations of DNR orders. POLST was designed to ensure that the full range of patient treatment preferences are honored throughout the health care system— outpatient, hospitalization, nursing home, and hospice. The brightly colored medical form begins by eliciting whether a patient would want cardiopulmonary resuscitation, but it then inquires about antibiotic use, hospitalizations, feeding tubes, and other options. Research shows that people want some treatments and not others, and these preferences need to be individualized. In one POLST study that involved people receiving hospice services, 99 percent of the patients did not want resuscitation efforts, three-quarters of the sample wanted at least topical antibiotics if they developed an infection, and one-tenth of the sample desired artificial nutrition if it was recommended.

There is also an intriguing debate taking place in the medical community about changing the nature of advance directives by "resetting the default." If this takes place, then the use of all life-prolonging measures would no longer be on the table, and it instead would be up to the physician with his or her professional knowledge and experience to determine which treatment options were appropriate. While this may seem like a step back in time to the era of paternalism, it is prompted by an appreciation of the flaw in overly relying on patient autonomy—especially when patients become demented or delirious and are unable to speak rationally. If the default is reset, decisions would still be reached through ongoing dialogues with patients and families, but these conversations would not start with the assumption that people want to be resuscitated. Instead, the purpose would be to elicit the patients' life values and goals and then use this information to arrive at a reasonable medical determination of the most correct option.

For example, in cases where no preferences had been previously stated, the discussion in the intensive care unit between the physician and family member of an incapacitated patient with Alzheimer's disease might begin with the doctor saying, "Research has shown that most people who consider the possibility that they might develop a severe dementia would prefer not to undergo aggressive life-prolonging treatments, such as cardiopulmonary resuscitation. We also know that even if these treatments are tried, they have only a small chance of reviving the heart and allowing the individual to improve sufficiently to ever leave the hospital and return home. Do you have any overriding reason to think that your loved one would insist upon such extraordinary treatments?"

Of course, resetting the default is itself a controversial idea. The concerns about palliative care expressed by Joyce Smith and others are not limited to hospitals, but they also apply to nursing facilities and residential hospices. Another social worker told me about being upset over the alacrity with which medications are commonly employed in palliative care. In the case of her ninety-one-year-old father, who was dying from lung cancer and residing in a hospice, she believed that use of a laxative, antihistamine, antianxiety medication, antipsychotic agent, and analgesic all contributed to sending him from a state of alertness to one of stupor and then to unconsciousness. Calling in a staff person to examine her comatose parent's breathing, she was horrified when the hospice nurse unexpectedly announced that he was in pain and then rapidly administered another dose of morphine. She didn't have time to voice an objection, and she thought, "I have just watched them kill my father!"

While medical personnel undoubtedly believe that they must rapidly ameliorate suffering, family members, like this social worker, may be resentful that greater effort and time are not expended on communication. Such families are forever left in doubt as to whether their loved one might have lived for an additional hour, day, or year; these lingering doubts largely could be put to rest by having everyone slow down, discuss the options, and take the necessary time to come to a decision. But this also requires us to prolong another human being's suffering—something that may be cruel and intolerable.

Although Olga and her allies see little that is ethical about the practice of palliative medicine, in reality those who practice it go to great lengths to stay within the established limits. Bioethics is the philosophical study of controversies engendered by advances in medicine and technology. Bioethicists explore the relationship between medical practice, theology, law, politics, and other intersecting bodies of knowledge. Like Baystate, every major medical center in the United States relies on bioethics to influence research initiatives through the oversight of institutional review boards and to assist with new clinical problems by means of bioethics consultation committees or services. In 1962, the Admissions and Policy Committee of the Seattle Artificial Kidney Center at the Swedish Hospital was probably the prototype for a hospital bioethics committee; its task was to determine which patients with renal failure should have access to the newly developed and very scarce dialysis machines. Ironically, today's committees are increasingly involved in decisions concerning not allocation but rather the withholding or withdrawing of life-prolonging treatments such as dialysis.

The majority of bioethicists in the United States have constructed a theoretical "bright line" to divide the different types of practices. On one side of the line, they accept the right of patients or families to forgo or stop life support. On the other side, they recognize that behaviors such as physician-assisted suicide that result in an "active shortening of the dying process" pose a more difficult situation.

While bioethicists also appreciate the contribution of religion to the debate, as a group they largely ascribe to a secular position. The palliative medicine philosophy in the United States has been formulated in concert with both the secular guidelines of bioethicists and several landmark Supreme Court decisions. The conventional view holds there is little or no moral/legal distinction between refraining from initiating life support and ceasing such treatments; furthermore, all life-prolonging medical therapies, including artificial nutrition and hydration, are roughly equivalent. American hospitals are generally liberal about withholding and withdrawal of life-sustaining treatments when these are the preferences of patients or their surrogates. Discontinuation of tube feeding is a more common situation in hospices and nursing homes. When religious considerations are important to patients and families, or when social situations are especially complex and ambiguous, consultations with ethics committees, psychiatrists, hospital chaplains, local religious leaders, and so on are encouraged and often helpful.

End-of-life decisions are greatly impacted by the religious affiliations of the patient, their loved ones, and medical staff. From what I can glean, Olga Vasquez has an especially deep reverence for life's sacredness, and both Kim and Amy raised the question of whether Olga's spiritual beliefs contributed to her accusation of murder. Olga is a member of the Assemblies of God, the largest Pentecostal denomination, with some 41 million members and adherents worldwide, and 2.5 million people who worship in more than 12,000 churches across the United States. Among its most prominent congregants have been Elvis Presley, the former Reagan administration interior secretary James Watt, and the notorious televangelists Jimmy Swaggart, and Jim and Tammy Faye Bakker. The 2008 Republican nominee for vice president, Sarah Palin, was a staunch member until 2002, when she moved to the nondenominational Wasilla Bible Church.

Rooted in a religious revival that began in the late nineteenth century or early twentieth century, the Assemblies of God believes there are sixteen individual doctrinal standards or truths; these deal with such matters as salvation, divine healing, baptism in the Holy Spirit, and the Second Coming of Christ. When contemporary issues such as abortion, racism, pornography, modern medicine, and suicide are not covered by doctrinal standards, they are discussed and measured alongside Scripture. The Church has position papers that articulate its official stance concerning these modern situations.

Some congregants practice speaking in tongues, miraculous healings, and a belief that their religion prohibits them from going to a doctor for any reason; other congregants include physicians and nurses who consider modern medicine to be a sensible means of healthy living. The most commonly held view is that one should rely on God for healing through prayer while also making use of medical interventions when appropriate. The church is firmly opposed to euthanasia and physician-assisted suicide, but is less clear about stopping life support treatment. Decisions about using or removing life-sustaining treatments are made by the individual member whenever possible after consultation with a Christian doctor and a respected spiritual leader.

The Assemblies of God literature states, "While weighing the biblical principle of respecting and preserving life, the Christian also takes comfort in an equal truth of joyously accepting our appointed time to begin eternal life with Christ after physical death."

The way that Catholicism reconciles these forces to each other is a bit different. Catholics make up a substantial proportion of Bay-state's population. Until quite recently, when I encountered Catholic patients considering dialysis termination I used to give a sigh of relief—our Catholic chaplains uniformly counseled people in a manner that was virtually identical with the hospital bioethicists. However, lately there seems to be a widening gap between Catholic practice and conservative Vatican theology—the former being more accepting of withdrawal while the latter expresses greater concern over violations of the sanctity of life. The greatest source of anxiety for conservative Catholic theologians seems to be that the momentum of palliative medicine may lead to an acceptance of physician-assisted dying and euthanasia.

Father John Paris, the Walsh Professor of Bioethics at Boston College, is representative of the more liberal viewpoint, and he has been an expert witness in many of the seminal right-to-life vs. right-to-die legal battles. When we met, Paris opined that for the past 450 years, Catholic moral teaching has differentiated between ordinary and extraordinary measures for preserving life. Originally expounded in the sixteenth century by Domingo Bañez and based on the views of Francisco de Vitoria, examples of extraordinary measures include those procedures that incur excessive pain, burden, or cost, and those that cannot demonstrate substantial benefit to the patient. In 1957, Pope Pius XII delivered what Paris considers to have been a remarkable address on the prolongation of life in the context of the hopelessly unconscious patient being kept alive with artificial respiration. In it His Holiness defined ordinary measures as being "means that do not involve any grave burden for oneself or another," and he concluded that after several days of unsuccessful attempts there was no moral obligation to keep a respirator going.

The Vatican's 1980 *Declaration on Euthanasia* substituted the terms *proportionate* and *disproportionate* for *ordinary* and *extraordinary*. It explicitly declared that if the burden of treatment would outweigh the benefits, refusal of treatment "is not the equivalent of suicide [or euthanasia]; on the contrary, it should be considered an acceptance of the human condition, or a wish to avoid the application of a medical procedure disproportionate to the results that can be expected, or a desire not to impose excessive expense on the family or the community." The declaration states that there is no need to undergo "forms of treatment that would only secure a precarious and burdensome prolongation of life."

According to Paris, a Jesuit scholar who delights in Socratic dialogues, "The broader context for this teaching is the Catholic understanding of life and death. . . . Life is thus a good, but not an absolute good. The absolute good for which we have been created is not this life, but the ultimate goal of union with God in eternal life. Death in this context is not an evil to be avoided at all cost, but the calling at some point in time of each of us back to God."

In our conversation, Paris made the further point that the purpose and goal of creation was to form relationships with others, and through those relationships one connects with God. When there is no more "capacity for ongoing relationship," he said, "the human task of striving for union with God through active love of neighbor has been completed."

However, following the experience of the late Pope John Paul II with Parkinson's disease, some members of the Catholic Church have begun to actively question this traditional view and are inquiring whether the withholding or withdrawal of tube feedings and perhaps other common palliative care practices should now be on the forbidden side of the bright line. In 2004, the pontiff declared, "The administration of water and food, even when provided by artificial means, always represents a natural means of preserving life, not a medical act . . . and as such is morally obligatory."

Paris would maintain that traditional Catholic teaching does not distinguish artificial nutrition and hydration from other medical life support treatments, and he has cited among other sources the guidelines issued by the Roman Catholic bishops of Texas in 1990. The bishops' analysis of patients who were in a persistent vegetative state was based on their interpretation of the Vatican's 1980 declaration, and it differs substantially from the view of those theologians who would describe the patients as "unconscious but nondying." The Texas bishops instead characterized them as "stricken with a lethal pathology which, without artificial nutrition and hydration, would lead to death."

Father Paris wrote:

> If there is evidence the now irreversibly comatose patient would not want to be maintained by artificial nutrition and fluids, these may be forgone or withdrawn. Such an action, in the bishops' understanding, is not abandoning the person. Rather, it is accepting the fact that the person has come to the end of his or her pilgrimage and should not be impeded from taking the final step.
>
> This subordination of physiological concerns to the patient's spiritual needs and obligations is the hallmark of authentic Catholic thinking.

In contrast to Christian martyrology, which sometimes encourages the believer to find value and edification in sacrifice, the Jewish tradition has only rarely considered suffering to be especially laudable or worth pursuing. Judaism, with its divisions into Orthodox, Conservative, and Reform types, would be complicated enough to parse into a unitary position on such matters, but according to the ethicist Dr. Y. Michael Barilan, one must also not underestimate that many Israeli Sephardic (Oriental) and Chassidic patients gravitate toward a form of mysticism; more like observant Muslims, these Jewish groups seek spiritual guidance from charismatic sages. Barilan quoted the son of one such patient as saying, "Every breath drawn by a Jew exerts positive influence in the divine court of Heaven."

Orthodox Judaism greatly values the preservation of life and with a few exceptions prohibits cessation of life-prolonging treatments (although again, it does not attribute any particular merit to human suffering). The extreme value of life is of importance in *halakhah*, the traditional Jewish rabbinic code of ethics and law, and an Orthodox colleague of mine has told me, "No matter how laudable the intention, an act that hastens the patient's death is equated with murder."

Culture and ethnicity are similarly important ingredients in these determinations. This was reinforced by a conversation that I had with my friend the Reverend Eugene Hodge, minister of the Apostolic Faith Church, who unequivocally stated, "Stopping life support treatment is the same as the taking of life." Picture a black, clean-shaven Abraham Lincoln with a French accent, and you have a reasonably accurate image of Hodge. Hodge's concept of death is tightly bound to unshakeable confidence that the Lord causes people to suffer for His own reasons and only He should be the one to dictate when life ends. Outraged over Terri Schiavo's fate, he emphatically announced, "Only God should be the one to determine who lives and who dies. That young woman could have had a miracle and become like everyone else. Once you open the door and remove feeding tubes and other treatments, pretty soon they will say that invalids or cripples should not be permitted to live."

Palliative medicine's emphasis—or overemphasis, depending on one's perspective—on quality of life being more important than quantity of life reflects a bias. To a large extent, it is a bias that is rooted in class and ethnicity. A nephrology colleague—a white woman who grew up in a middle-class family—discovered this during a conversation with one of her severely debilitated black patients who had persistently struggled to survive an impoverished background. The physician broached the subject of stopping renal replacement therapy by talking about the obvious erosion in his "quality of life." However, when it was apparent that death would be the consequence of dialysis discontinuation, the patient challenged his physician's logic and replied somewhat irritably, "But dead is no quality of life!"

Statistics demonstrate that in our country African Americans and Hispanic Americans with kidney failure are only about half as likely as Caucasians to agree with discontinuation of their dialysis. Aside from the possibility of a greater role for religious beliefs in many non-white groups, another factor in the decision to refrain from stopping renal replacement therapy is a deeply rooted distrust of the medical community and the establishment. People from minority groups remember historically flagrant violations of research ethics—such as the Tuskegee Syphilis Study, where a group of African American men were not given informed consent, accurately told their diagnosis, or provided with penicillin when it became available—as well as other experiences of blatant racism in the provision of medical care. Discussions are invariably more complicated when patients or the hospital ancillary staff are nonwhite and interact with physicians and nurses who may not share their religious and cultural beliefs and their ethnicity.

In the end, Olga and her allies at Baystate are merely a few of the many foot soldiers involved in a campaign that is questioning the tenets of palliative medicine. Their beliefs as to what constitutes murder in a medical setting are important, but they are not the leaders in shaping the overall agenda that is confronting palliative care. To meet and understand the movement's generals, it's necessary to look back on the case that thrust this issue into the international spotlight.

10

The Coalition

I t's significant that Rosemarie Doherty's death and Olga's accusation of murder predated the swirling controversy surrounding the last days of Terri Schiavo, as it was the Schiavo case that gave a voice to accusations like Olga's, which until then had been playing out on a much smaller scale. While the sociopolitical war over how we die and the opposition to many of modern palliative medicine's end-of-life care practices did not begin in America with Terri Schiavo, her fractious case stands out because of the frenzy of media and public involvement it generated and the way in which it was repeatedly punctuated by accusations of murder. Though the roots of the controversy are plainly evident in earlier Supreme Court cases, it has been the acrimony and persistent efforts of Terri Schiavo's family—her husband, Michael Schiavo, who faced off against her parents and siblings, the Schindlers—that brought the conflict into sharp focus and makes one appreciate the take-no-prisoners seriousness of this fight.

The Schindler family's views are both vitalist (every human life needs to be maintained as long as possible and at any cost and without regard to quality of life) and theist (it is a mortal sin to attempt to assume control over the manner of one's death). According to this latter stance, any efforts to do so must be seen not just as overweening efforts to "play God" but as acts that profoundly diminish the sacredness of life.

These sentiments were glaringly illustrated in the report of the court-appointed guardian ad litem, Dr. Jay Wolfson, to Florida governor Jeb Bush:

> Throughout the course of the litigation, deposition and trial testimony . . . members of the Schindler family voiced the disturbing belief that they would keep Theresa alive at any and all costs. . . . [There was] agreement by the family members that in the event Theresa should contract diabetes and subsequent gangrene in each of her limbs, they would agree to amputate each limb, and would then, were she to be diagnosed with heart disease, perform open heart surgery.

During the Supreme Court battles in the 1970s and 1980s that were prompted by the cases of Karen Ann Quinlan and Nancy Cruzan, opposition to withdrawing life support included a coalition-of-convenience composed of conservative government officials, religious leaders, some medical authorities, and others. During and after Terri's death, the Schindler family was instrumental in approaching pro-life activists, disability rights representatives, and other groups that revived and reinvigorated the coalition. As will be discussed, the coalition has become an international phenomenon, and although it lacks a formal name, its proponents gather under a banner that proclaims the sanctity of life and their opposition to all forms of assisted dying. The leaders of this coalition maintain that the Schindler family beliefs represent a minority opinion; however, there is no denying that Terri Schiavo is the coalition's Joan of Arc. The vitalist and theist positions may not be uniformly held by the other constituents of the coalition, but they are emotionally central to the movement, and they fuel many of its most potent, rhetorical sound bites.

When Terri Schiavo's life came to an end on March 31, 2005, Florida governor Jeb Bush issued a condolence statement: "Many across our state and around the world are deeply grieved by the way Terri died. I feel that grief very sharply. . . . I remain convinced, however, that Terri's death is a window through which we can see the many issues left unresolved in our families and in our society."

Shortly after her death, strong words were heard from the Vatican, where Renato Cardinal Martino, prefect for the Pontifical Council for Justice and Peace, proclaimed, "It is a victory of the culture of death over life. This is not a natural death, it is an imposed death." The cardinal angrily condemned the circumstances as being "nothing else but murder."

Cardinal Martino was only one of many people who brought up the subject of murder. This began with charges from Terri Schiavo's birth family, the Schindlers, that her husband, Michael, was responsible for the original brain injury. The Schindlers claimed Michael either struck Terri in the midst of an argument or neglected to expeditiously summon medical assistance when she lost consciousness from a cardiac arrhythmia.

Mark Fuhrman has written a book about Terri Schiavo. The former Los Angeles homicide detective is chiefly known for finding the "bloody glove" at the scene of the murder of Nicole Brown Simpson and Ronald Goldman, along with a matching glove in O. J. Simpson's yard. He began investigating and writing about Terri with the full cooperation of the Schindler family and at the urging of conservative political commentator Sean Hannity. In his book, Fuhrman maintained that the Schindler family begged law enforcement agencies to examine charges that Michael Schiavo abused their daughter. In 2002, they petitioned the court and submitted evidence, including a bone scan, alleging domestic violence. Fuhrman stated, "It is abundantly clear that as soon as he received the medical malpractice award, Michael Schiavo began a relentless campaign to end Terri's life"; later he wrote, "The state of Florida committed a legal homicide upon Terri Schiavo . . . on the direction of Michael Schiavo." However, at the end of his exposition Furman was reluctantly forced to conclude that he had discovered no grounds for criminal prosecution.

The Schindlers remained adamant that at the very least Michael had committed perjury and killed their daughter by falsely testifying that Terri had said she would prefer to die rather than live in a coma. The issue of criminal accusations resurfaced during a tense judicial hearing when David Gibbs, the Schindlers' lawyer and president of the Christian Law Association, publicly castigated Michael Schiavo for being a murderer, and the presiding judge, James Whittemore, promptly admonished Gibbs for evoking "the emotional aspect of this case."

However, Michael Schiavo was not the only person to be branded a criminal during Terri's saga. When hundreds of enthusiastic protestors descended upon Hospice House Woodside (part of the Hospice of the Florida Suncoast) in Pinellas County, Florida, they held up signs denouncing "Hospice Auschwitz" and shouted "Murderers!" at the hospice staff. It made no difference whatsoever that this institution had won the prestigious Circle of Life Award from the American Hospital Association and was committed to supporting both the Schiavo and Schindler families. To the demonstrators, the staff was actively participating in a crime. Furthermore, the allegations did not solely target the hospice workers. The protestors also waved handwritten placards in front of the cameras comparing Florida probate judge George Greer to Adolf Hitler. They also angrily accused him of being a murderer. Judge Greer's purported criminal behavior consisted of upholding the law that granted Michael Schiavo—the spouse—the right to make decisions about the withdrawal of his wife's feeding tube. On Easter Sunday, the neighborhood of the hospice was repeatedly punctuated by loud chants from the demonstrators proclaiming that the police, too, were Nazis.

The protestors at Hospice House Woodside were among the many recruits and enlistees of the coalition. Especially prominent in the coalition are members of the pro-life movement, whose involvement began, according to Randall Terry, founder of the anti-abortion group Operation Rescue, when "the [Schindler] family asked me to create an avalanche of publicity on behalf of Terri." Randall Terry and his supporters responded by organizing prayer vigils outside the hospice where Terri Schiavo lived for the last three years of her life, and they made frequent and repeated radio and television appearances.

Religious conservatives played a pivotal role in this case by coming out in force with an arsenal of prayer vigils, Christian radio broadcasts, and thousands of e-mail messages targeting state and federal lawmakers. Although at the present time they would strategically like to avoid the limelight, fundamentalist, evangelical, and other religious groups have remained the central power of the coalition.

Disability rights groups are divided by these issues; a total of fourteen national organizations filed friend-of-the-court briefs in support of the position held by the Schindler family. The disability organizations that are part of the coalition are greatly concerned about further devaluation of life for the disabled.

One of these organizations, the colorfully named Not Dead Yet, originally began in reaction to the 1996 acquittal of Dr. Jack Kevorkian following two cases of assisted suicide. A spokesperson for the group stated, "Initially the focus of . . . heavily-funded 'end of life' [palliative medicine] advocacy groups . . . was on making services better for people in the last stages of a terminal illness. Over the last couple of years . . . there's been an alarming broadening of their agenda . . . to extend the definition of 'terminal' to include people with incurable disabilities and chronic conditions whose future lifespan can be measured in years."

Another activist wrote that the judicial sanctioning of decisions such as the removal of Terri Schiavo's feeding tube "moves America back to the days when the sterilization and elimination of people with disabilities did not merely reflect private prejudices but were embraced as the law of the land."

The disability rights organizers I spoke with were pleased to have a forum to express their platform, but many were thoroughly annoyed at being misquoted by the media and manipulated for political reasons by both sides of the dispute.

It was the withdrawal of artificial nutrition and hydration that especially drew the ire of the Schindler family and remains anathema to the coalition. In 2004, Pope John Paul II stated that health care providers are morally bound to provide food and water to patients in persistent vegetative states. In 2005–2006, the various right-to-life groups tried, but were unsuccessful, to get more than twenty-five separate state legislatures to pass a model bill titled the Starvation and Dehydration of Persons with Disabilities Prevention Act. Wisconsin governor James Doyle vetoed one such statute, explaining that it should be called the "unconscionable clause . . . because it puts a doctor's political views ahead of the best interests of patients."

However, in June 2008, the Arizona state legislature and governor passed Jesse's Law, which prohibited surrogate decision makers from readily denying food and fluids to an incapacitated patient. The measure was inspired by the case of Jesse Ramirez, a postal worker and Gulf War veteran who sustained a traumatic brain injury in a car accident and whose physicians withdrew a feeding tube and transferred him to a hospice. Like Terri Schiavo, the family was split in a legal contest between the spouse (who supported withdrawal of the artificial nutrition and hydration) and the patient's sisters and mother (who were opposed). The Alliance Defense Fund, a Christian legal group, was instrumental in having the courts order the artificial tube feeding reinstituted while legal matters were being resolved. Unlike Terri Schiavo, Ramirez was not in a persistent vegetative state, and he emerged from his coma. This dramatic turn of events promptly led to passage of the law, which has been widely celebrated by the pro-life community.

The coalition is correct in its assumption that physicians can make errors in distinguishing persistent vegetative states from other conditions that follow brain injury, but this does not alter the prevailing opinion in medicine about feeding tubes, which is that they are a medical treatment valuable only for sustaining life under specific circumstances. Use of the tubes is certainly not the same as eating or feeding, as there is none of the touch, smell, and taste normally associated with food. One of my closest colleagues at Baystate, a nurse named Dianne Baker, told me that artificial feedings will remain in her opinion a medical procedure until the day we are able to squeeze a cheeseburger down the tube. She has written an advance directive for herself that explicitly forbids their use under any circumstances. The late Reverend Chuck Meyer of Texas similarly concluded, "It is not a feeding tube unless you think you can get an enchilada down that sucker. You think it's food? You eat it!"

Not everyone who has problems with palliative care takes the same stance as the Schindlers. Alex Schadenberg is the executive director of the Euthanasia Prevention Coalition and the organizer of the Toronto symposium where I met Bobby Schindler. An indefatigable young man who resides in the small community of London, Ontario, he laughed merrily when describing his physical appearance to me as "completely ordinary." Alex studied history at university and then worked over the next five years for the Catholic Church, running the Family Life and Youth Ministry Office. He began his position at the Euthanasia Prevention Coalition in July 1999 as a reaction to his personal horror over the widespread public and judicial support of Robert Latimer.

Latimer was a farmer from Saskatchewan who killed his twelve-year-old daughter, Tracy, on October 24, 1993. Tracy had severe cerebral palsy, and at the time of her death she was a forty-pound quadriplegic functioning at the level of a three-month-old. Throughout her life she required extensive medical and surgical care and was unable to walk, talk, or feed herself. Although she responded to affection, and according to some observers was capable of smiling, Tracy was in constant, excruciating pain when Robert Latimer killed her in his pickup truck by carbon monoxide poisoning (he ran a hose into the vehicle's window).

Found guilty in 1994 of second-degree murder, he stood trial again in 1997, with the jury arriving at the same verdict but recommending that he become eligible for parole after a year—even though the minimum sentence for second-degree murder is twenty-five years with no chance of parole for ten years. Justice Ted Noble granted Latimer a constitutional exemption from the minimum sentence, stating that Tracy Latimer's murder was a "rare act of homicide . . . committed for caring and altruistic reasons." He went on to say, "That is why for want of a better term this is called compassionate homicide." The case was appealed and it journeyed back and forth to the Canadian Supreme Court before Latimer was ultimately released for day parole on March 13, 2008. A substantial portion of the Canadian public felt that he never should have been imprisoned.

One of Alex Schadenberg's six children has autism, and consequently he has a long-standing and intimate connection with the disability community. His reaction to the Latimer case was to ask, "If Tracy's life lacks value to that many Canadians, then what about other vulnerable people?"

Alex, like all the coalition spokespeople that I encountered, was very careful and showed great subtlety and sensitivity in his critique of palliative medicine. He explained that overall he supports the specialty and its efforts—with the exception of a few individual palliative care physicians and hospice staff who have "crossed a line between killing and letting die." To this end, he is worried these few palliative physicians are a bad influence and that the field is evolving and changing in unacceptable ways. Otherwise, Alex was comfortable about palliative care involving vigorous amelioration of symptoms—especially when directed at the dying—and he could agree with the principle of double effect and the recognition that some treatments may hasten the dying process. He was even accepting of the occasional need for palliative sedation—a practice that I will discuss more fully later, but which entails the induction of a coma for unremittingly severe suffering.

However, Alex became visibly more emotional in the interview when talking about the threshold for terminating artificial hydration and nutrition. He was obviously disquieted by hospice programs that have explicit written policies with preferences toward discontinuing hydration. His belief is that the dying should be fed by hand and spoon if possible, and if they are unable to swallow, then they should continue to be provided with at least small amounts of liquids by tubes and other mechanical means. Alex and the other leaders in the coalition were especially outraged and indignant at the thought of nutrition, hydration, and similar life-prolonging treatments being terminated or withheld from individuals who were not actively dying. This is the flashpoint where Alex forcefully evoked the platform of disability rights; he maintained it is pure hubris when families and society believe they can judge the value or worth of human life.

Alex mentioned to me his personal bête noir, the philosopher Peter Singer, who he said holds the position that when the demented or brain-dead largely lack consciousness and awareness of others, then they are no longer in the same category as sentient human beings. Alex strongly rejects that position, and he raised the specter that it could readily lead in the future to Alzheimer's units being emptied out by staff who decided to cut costs and terminate the nutrition and hydration of their patients. His scenario is not dissimilar to ex-governor Sarah Palin's rhetorical flourish in the 2009 American health care debate about physician "death panels" that would judge Down syndrome children—like her son—to be unworthy of life. While such ideas may seem ludicrous, they are not entirely fanciful if one recalls the medical policies in Nazi Germany during World War II.

About Terri Schiavo, Alex told me that if she had contracted cancer after incurring the severe brain damage, he might have felt differently about what transpired (this is in contrast to the view expressed by her family). But, he explained, she was not dying when the feeding tube was removed, and "the intention was to cause her death, the action was to remove fluids, and in the end she died of dehydration. . . . That," he unambivalently declared, "is euthanasia without a doubt."

According to Alex, dehydration of people who are otherwise physiologically stable is even crueler than injecting them, because it drags out the killing process over days or weeks. He is aware of the research and clinical experiences that dispute his conclusion, and he explained that most of those studies of death by dehydration involve patients with metastatic cancer—which he feels alters the body in a way that is different from other conditions.

Furthermore, Alex expressed concerns over the exaggerated importance assigned to autonomy when seemingly competent individuals requested or demanded that life-supportive measures—feeding tubes, dialysis, ventilators, or other treatments—be stopped. His fear is that people make these requests when they are mentally incapacitated by depression and troubled by disproportionate concerns over the burden of their illness on loved ones. He would like such individuals to explore the following questions before taking any action: "Am I nearing death or am I not nearing death? Am I causing death or am I not causing death? Are my social needs and emotional needs being met the same as my physical needs?"

Later in the day, Dr. William Toffler, the national director of Physicians for Compassionate Care, similarly remarked, "When a patient says, 'I might as well be dead,' they may really be saying, 'No one cares about me.'"

Alex and I gingerly talked about the Catholic Church. He did not wish to falsely come across as a professional theologian, but he said, "Allowing natural death to occur is not a problem; killing someone is a problem. . . . If you kill people, then you don't fit within the Church. If you are promoting killing, you are obviously outside of what the Church is teaching."

Alex and I revisited the case of Piergiorgio Welby, an Italian man with muscular dystrophy who in 2007 was denied a Catholic funeral after his ventilator was disconnected by a physician. Welby had poignantly written about his wish to die, and most of the accounts I have seen state that the Vatican declared his death to be from euthanasia. Several Italian politicians actively sought to indict the physician for murder. Alex explained to me, "Welby died very quickly after his ventilator was stopped. This was not euthanasia . . . it was a natural death. While it is true that he was denied a Catholic funeral, when you carefully read the documents it is very clear that this had nothing to do with how he died. It was [on account of him being] a euthanasia campaigner—he was considered someone outside of the church. . . . He was excommunicated. In fact, he excommunicated himself. . . . If you do euthanasia or are a celebrity who promotes euthanasia, you have excommunicated yourself. That's how the church documents are written. . . . It would have been a scandal for them to have given him a Catholic funeral."

• • •

Dr. Peter Saunders and I met over breakfast before the Toronto anti-euthanasia meeting started. Despite his obvious jet lag from having just flown in from Heathrow, he was handsome, elegantly appointed, and irrepressibly eager to describe his work in Great Britain. Peter is the general secretary of the Christian Medical Fellowship, and he is also the campaign director for Care Not Killing (CNK). The Christian Medical Fellowship is one of the core members of CNK, which is an alliance of forty-eight organizations, including palliative care, professional, faith, and disability rights groups. CNK appears to be the British equivalent of Alex Schadenberg's Euthanasia Prevention Coalition.

Peter explained, "The groups in the CNK come from a very broad background but are united in our commitment to promoting good palliative care, and to opposing any changes in law that would allow euthanasia or assisted suicide. The CNK came together to oppose a bill in 2005 that would have permitted physician-assisted suicide, and in the process not only defeated it but also turned around the position of the British Medical Association and the Royal College of Physicians, and the Royal College of General Practitioners." All of these organizations had previously been neutral on the subject of euthanasia, and Peter was pleased that they each have now stated their opposition to this practice.

CNK expects that there will soon be another attempt to change the law in Scotland—"another Oregon-type bill"—and the organization is remaining vigilant. "We have to be wary," Peter told me, "of back-door routes, like the withdrawal-of-treatment issues." This statement initially floored me, because Peter had begun the interview by seemingly declaring his wholehearted support of palliative medicine. He went on to say, "The other back-door route that we are wary of is a watering down of the homicide laws leading to less severe penalties to those who cross the boundaries." He explained that in the Netherlands, euthanasia had been legally sanctioned long before it was officially legalized—if Dutch doctors followed a set of guidelines, they wouldn't be prosecuted. He cited the case of Dr. Geertruida Postma, who killed her mother and then received a relatively minor reprimand from the courts: "When laws aren't enforced, people will transgress." He said that acceptance by the Dutch medical society and "more liberal views about withdrawal of treatment combined to produce the cocktail of euthanasia being legally sanctioned and then finally legalized in 2002."

I was intrigued by his tough law-and-order stand and his allusions to treatment discontinuation, and he explained, "CNK has a clear view on euthanasia and physician-assisted suicide, but we haven't always been able to arrive at a consensus view that has pleased every one of our member organizations on the withdrawal and withholding issue. . . . You can imagine that keeping together a group as diverse as the CNK is quite a challenge. . . . This is probably the issue for which there has been the most tension . . . there are some groups within CNK that I would call vitalists. [They] have great difficulties with the withdrawal of any life-sustaining treatment. There are others who have a much more liberal perspective. I am, personally, probably somewhere in the middle."

Echoing Alex Schadenberg, Peter's view was that "killing people is always wrong; letting people die sometimes is and sometimes isn't—depending on the situation. I think the key issues are, first of all, what is one's intention in withdrawing treatment—is the intention that the patient die, or is one's intention to remove a treatment which is worse than the disease? And I think the other key question is what actually kills the patient—is it the disease or is it the drug that is being given or the deprivation of food and fluid? I don't believe we are obliged to give every available treatment to every patient no matter what the circumstances. I believe it is good medical practice to withdraw and withhold treatments when the burden of those treatments outweighs the benefits in a patient who is approaching death—where death is inevitable and beyond our power to stop it—while recognizing that on occasions this may have the effect of shortening life. In other words, there is a time to die, but I don't think there is a good time to be killed."

CNK has been troubled by Britain's Mental Capacity Act, and Peter explained, "We oppose legally binding advance directives. We believe withdrawal of treatment decisions should always be left to health care professionals and patients to make together. We have reservations about advance directives, in the sense we believe that patients don't always make them in the best and informed way. We also know that many patients would want their doctors to overrule the advance directives if they felt them to be clinically inappropriate. So we're against the whole idea of forcing doctors to act with a hand tied behind their backs—in effect, against their better judgment.

"People," he told me, "fall into two camps. We have a situation of what we may call 'furious opposites.' The one opposite is the overaggressive, inappropriate, meddlesome treatment of patients who are nearing death—such that the treatment becomes worse than the disease. [That engenders in patients and families] what I would call the furious opposite of overtreatment. At the other end is the furious opposite of negligence or undertreatment. It is a very rare person who can balance those two furious opposites and walk the path equally between them. Most people tend to veer to one side or the other—in other words, they fear overtreatment more than they fear undertreatment or vice versa." He concluded with the apt observation that people's positions tend to be colored and formed by their own life-and-death experiences with loved ones.

I asked Peter about his reference to vitalist members of CNK, and he described them as being more inclined toward medical interventions and preferring "overtreatment rather than undertreatment."

He identified two of the alliance's pro-life organizations, ALERT and the Society for the Protection of the Unborn Child (SPUC), and said, "They are now quite interested in end-of-life issues as well. They are probably the two groups that oppose the Mental Capacity Bill the most. I would say that the Christian Medical Fellowship is middle-of-the-road, as are some of the other pro-life groups. Probably the Association for Palliative Medicine would be . . . more afraid of overtreatment." The last turns out to be the only non-faith-based professional group in the Care Not Killing alliance.

Listening to voices such as these, it's easy to appreciate Bobby Schindler's and many others' negative reaction to the practice of palliative medicine in our hospitals. And while I disagree with their position, I can to some extent—especially on an emotional level—relate to their viewpoint when they maintain that we are killing our patients or permitting them to kill themselves by truncating treatment.

Indeed, this emotional response of a murder accusation even makes a strange degree of sense—at least as a coping mechanism. My colleague Dr. Mike Germain maintains that the psychological explanation for both end-of-life accusations of murder and current efforts to criminalize end-of-life medical practices are linked to humankind's pervasive anxiety and fear about death. Woody Allen is hardly alone in his sentiment that "rather than live on in the hearts and minds of my fellow man, I'd prefer to live on in my apartment." When a loved one dies, a part of us also dies. It is not uncommon for grief—or, for that matter, anticipatory grief—to contain varying degrees of despair, anxiety, guilt, rage, and sadness. It is not unusual for these feelings to be displaced onto other individuals or institutions. Civil malpractice litigation commonly ensues if one holds a physician or hospital responsible. But murder accusations take matters to another level.

Some of these allegations may arise from preexisting and petty conflicts between individuals that are then enacted in courtrooms. However, a more generous explanation is that there are many people, including the Schindler family, who sincerely believe suffering is part of a religious design in which humans should not intervene—especially if that intervention hastens dying. There are other people who hold a secular form of vitalism that stems from the idea that life in whatever condition is precious, and death is the ultimate enemy. Many of these individuals are increasingly squaring off with medical professionals who affirm the philosophy of palliative medicine and seek the amelioration of symptoms in the final days of life. From the perspective of those whose highest regard is accorded to the sanctity of life, it is understandable that a physician or nurse who believes otherwise and accelerates dying is improperly practicing medicine and is guilty of killing.

In my conversations with members of the coalition it is clear that some leaders, such as Bobby Schindler, conflate and oppose euthanasia, physician-assisted dying, and the philosophical underpinning of palliative medicine. Other leaders, such as Dr. Bob Orr from Vermont, who also spoke at the Toronto meeting, recognize distinctions and are frank advocates of palliative care. It is an oversimplification to conclude that the coalition is uniformly opposed to the use of palliative medicine in end-of-life care, but it is accurate to say there is widespread skepticism and sometimes outright rejection of specific practices, such as the withdrawal of artificial nutrition and hydration or the aggressive use of opiates.

It does not appear to me that there has been a concerted effort on the part of the coalition to target clinicians with criminal accusations. Rather, I believe that when health care professionals are accused of murder and investigated, this is a by-product of the philosophical controversy—a conflict that has been inflamed in recent years in the United States by the widespread reliance on planned deaths following withdrawal and withholding of life support treatment and the provision of hospice services. The ascendancy of palliative medicine practices in our society and suspicions about the motives of individual nurses and doctors have combined to create an acrimonious, litigious, and dangerous atmosphere for health care practitioners. Added to this is the undeniable reality that there actually are deviant medical personnel who murder patients.

11

Serial Medical Killers

A s it turned out, Amy, Kim, and Olga were taking care of Rosie at a moment when all of western Massachusetts was preoccupied with the trial of suspected serial medical killer Kristen Gilbert. Four years earlier, three nurses from the Veterans Affairs Medical Center in Northampton had come forward to report concerns about a colleague. Their suspicions were fueled by an unusual increase in cardiac deaths at their hospital and an unexplained decrease in the supply of a medication, epinephrine. Attention focused on Kristen Gilbert, a registered nurse. Because of the unexpectedly high mortality rate occurring during her shifts, Kristen was jokingly referred to by her colleagues as the "Angel of Death." Police detectives, however, do not believe in coincidences, and they thought it was highly suspicious that during a three-year period covering Ward C, Gilbert had also reported eight separate fires and received an award for extinguishing one of them.

The investigation ultimately revolved around the deaths of seven patients. Assistant U.S. attorney Ariane Vuono explained to the jury that the hospitalized veterans had "protected our country during war and peace. They were vulnerable, due to their physical and mental illnesses. Some were seriously ill. And some had no family. And because of that, ladies and gentlemen, they were the perfect victims."

The trial was a sensational event in our part of Massachusetts, a place where the police blotter of the local newspaper, the *Daily Hampshire Gazette*, is more likely to report a mailbox vandalized, a car break-in, or University of Massachusetts students getting arrested for rowdiness at a house party. In 1998, Kristen Gilbert was thirty years old when she was indicted for murdering four patients and attempting to murder three others by injecting them with a heart stimulant. All the victims were elderly men, but only one was terminally ill; the government's position was very clear that it never considered these to be mercy killings. Although the Commonwealth has a long-held prohibition against capital punishment, the alleged criminal acts were committed on federal soil—the Veterans Affairs Medical Center—meaning that if found guilty, Kristen Gilbert could be sentenced to death.

Given the stakes, Gilbert was fortunate to be assigned to an especially sagacious federal judge, Michael Ponsor. Ponsor—the same judge who was quoted earlier regarding his friendship with the shoeshine man, Benjamin Babcock—brought to the case seventeen years of experience on the bench. Some fifteen hundred jury summonses were sent out, and because the courthouse was insufficiently large enough to contain the juror pool, eight hundred potential jurors heard preliminary instructions and completed a fifteen-page questionnaire at nearby Springfield's Symphony Hall.

Eventually twelve jurors and six alternates came to the Northampton courthouse and heard the case presented from November 20, 2000, to March 26, 2001. Three extraordinarily dedicated local lawyers constituted Gilbert's defense team, assisted by three investigators, two toxicologists, a pathologist, two cardiologists, a nursing consultant, a jury consultant, a venue analyst, two mitigation specialists, a statistician, a neuropsychologist, a behavioral psychologist, a psychiatrist, an endocrinologist, and a paralegal. By the conclusion of the trial, the defense had spent thousands of hours and $1.6 million of public funds preparing the litigation. It is safe to assume that at least the same amount was expended on the Commonwealth of Massachusetts' case.

The prosecution charged that Kristen Gilbert had not only betrayed her responsibilities as a nurse and turned her back on a patriotic duty to care for veterans but also committed adultery. It was widely known at the VA Medical Center that she had engaged in a torrid affair with a hospital security guard. According to the district attorney, sex was the salacious key to understanding Gilbert's criminal behavior. Precipitating cardiac arrests and declaring codes brought multiple hospital personnel, including her paramour, rushing to patient bedsides, where they observed Kristen valiantly attempting to save lives. In the words of the prosecutor, Kristen Gilbert was a "code bug," the same way an arsonist is a "fire bug." (For that matter, although there was evidence suggesting she actually was a fire-setter, she was never prosecuted on any charges of arson.)

On March 14, 2001, Gilbert was convicted in federal court of three counts of capital murder, one count of second-degree murder, and two counts of attempted murder. At the conclusion of the penalty phase, the jury was deadlocked; Judge Ponsor sentenced her to mandatory life imprisonment without possibility of release. Gilbert is now serving four consecutive life sentences at a maximum-security prison in Fort Worth, Texas, where she will eventually die.

After hearing from the nurses at my hospital, I wanted to learn more about the circumstances of the case against Kristen Gilbert. I suspected that when the district attorney began vigorously pursuing the investigation of the Baystate staff he was wondering whether there was a hidden epidemic of medical homicides taking place in western Massachusetts hospitals.

Kristen Gilbert's guilt seems pretty clear; however, I decided to look a bit deeper into the case and contacted a colleague named Dr. Jim Kirchhoffer, who had testified as an expert witness for the defense team. Now, a day before our scheduled meeting, he was pacing in the hospital corridor outside my office. Jim, an intense and devoted cardiologist—known to his many friends as "Captain Kirk," after the indomitable commander of the starship *Enterprise*—was not comfortable waiting. We had set an appointment for Tuesday afternoon, but it was now Monday and he had clearly gotten the dates mixed up. Standing outside my door, Captain Kirk reminded me of a muscle car gunning its engine before the start of a drag race.

When I was finally free to see him in my office, Jim leaned forward in his thickly cushioned Morris chair and rapidly reviewed the facts of the case. "I had no clue as to what I was getting myself into when I was called to be an expert witness," he told me. "And, as things developed, it changed the lives of everyone who was in it." He went on to explain that the legal team and its consultants were passionately involved in the defense; after the unfortunate conclusion of the case one attorney left the practice of law, a partnership fell apart, and several marriages foundered.

"Kristen Gilbert was charged with hastening the demise of a bunch of Veterans Administration patients by injecting them with epinephrine. I thought, 'That is crazy! Epinephrine is a life-saving drug. Epinephrine will kill people when pigs fly out of my butt.'"

Grinning at me, he said, "In fact, I made this last statement to the defense team so many times that they developed an entirely unwarranted fear that I was going to repeat it while testifying on the stand."

For the first time, I understood why Jim was wearing a tiny gold charm of a pig with wings conspicuously glued to his Baystate identification badge. It was given to him by Gilbert's attorneys, and it openly demonstrated his continuing attachment to the case. Since he was convinced that Kristen Gilbert was innocent of murder, Jim could not understand why she had been successfully prosecuted. He found the district attorney's theory that Kristen had repeatedly induced cardiac arrhythmias in patients so that she could be seen heroically attempting to resuscitate them—much like a fireman who commits arson—to be entirely farfetched and unlikely.

He claimed bitterly that the particular VA hospital where she worked at the time was "a snake pit," staffed by a number of people who drank, used recreational drugs, and slipped off to the parking lot for extramarital sex. In Jim's view, Gilbert may not have been entirely chaste or morally perfect, but she was seriously damaged by immaterial testimony alleging she was an adulterous mother. He also surmised that the assistant U.S. attorney who prosecuted the high-profile case may have seen an opportunity for professional advancement and publicity. In fact, four of the criminal investigators, special agents, and prosecuting attorneys in the Gilbert case were subsequent recipients of national awards from the Federal Law Enforcement Officers Association.

Several months before I met with Jim, Amy Gleason had approached him, and he was aware of her situation. I asked for his thoughts, and he replied, "It is an arrow directly through the heart. Here are Baystate nurses who are trying to save these poor suffering people, and some little twerp of a TA says, 'You are not trying to ease their suffering. You are trying to kill them!' It is unbelievable."

Having never been personally accused of a criminal act—having never received so much as a speeding ticket—I am unlikely to think about crime other than when I occasionally read mystery novels or encounter newspaper stories. Accordingly, I was dumbfounded and initially felt the same indignation as Jim when Olga Vasquez told me that Kim Hoy was a murderer. It seemed outrageous that a fellow physician or nurse could engage in homicide or require a criminal investigation. Although it was understandable that Olga and the district attorney might be preoccupied with thoughts of Kristen Gilbert, I considered that case to be anomalous.

The truth, however, is always more complicated, and just a cursory search into this issue of medical serial killers demonstrates how serious a problem it really is. Gilbert is hardly the sole medical serial murderer; indeed, a brief Internet search turned up a considerable number of nurses, including Rhea Henson, Terri Rachals, Donald Harvey, Orville Lynn Majors, and Robert Diaz. Diaz was convicted of murdering eleven patients and currently resides on death row at San Quentin, where he is sentenced to die in the gas chamber.

I do not wish to trivialize or make light of the heinous things I am about to discuss. But the only way I can compartmentalize the following series of crimes is to take a walk down an imaginary prison corridor and briefly "meet" some of the medical personnel who are inmates.

In the first cell we might find the now forty-nine-year-old former nurse Charles Cullen. In December 2003, he was arrested and charged with murdering a Roman Catholic clergyman and attempting to kill another patient at Somerset Medical Center. Cullen stunned investigators by announcing in court he was prepared to plead guilty, and that during his sixteen-year nursing career he had ended the lives of forty other patients in Pennsylvania and New Jersey. He confessed to administering overdoses of insulin and digoxin—medications used for diabetes and cardiac disorders—ostensibly to alleviate pain and suffering. Although Cullen killed patients who were terminally ill, he also killed others who were recuperating from their illnesses. His victims ranged in age from thirty-eight to eighty-nine.

Cullen has a long history of mental illness and attempted suicide at least three times; he was psychiatrically hospitalized on four occasions. According to the *New York Times*, he had a record of abusing dogs and cats, and he also stalked a fellow nurse. At his sentencing in April 2004, he agreed to a plea bargain of thirteen life sentences in prison, which makes him ineligible for parole for the next 127 years.

As we walk along our imaginary cell block, we next encounter Vicky Dawn Jackson. She is suspected of killing twenty-three patients at Nocona General Hospital in Texas using injections of mivacurium chloride, a paralytic agent that stops people from breathing. In 2001, the hospital alerted authorities that several bottles of the medication were missing, and this led to an investigation that implicated Jackson.

Genene Jones is in a cell, and her case makes the point that sometimes the victims are children. In 1984, this pediatric nurse was sentenced to ninety-nine years in prison, and in a subsequent trial she was sentenced to an additional sixty years. Jones is suspected of having murdered approximately fifty infants.

This prison tour needs to be international in scope, as cases of murderous health care personnel have occurred in many countries other than the United States. In February 2003, Christine Malèvre, a French nurse, was sentenced to ten years in prison for the deaths of six hospital patients. She is charged with murdering patients at a pulmonary hospital in Mantes-la-Jolie, near Paris. Malèvre wrote a controversial book about her case and initially said that she had helped about thirty terminally ill people to end their lives.

Hungary warrants a visit. In 2001, a nurse there, Timea Faludi, was found guilty of attempted murder. Despite a confession that she actually gave lethal injections to some forty elderly patients at the Gyula Nyiro hospital, police have only been able to collect solid evidence in eight instances.

A trip to Japan is also necessary in order to meet the nurse Daisuke Mori. Accused of fatally administering muscle relaxants instead of prescribed medications, he has been nicknamed by his colleagues "Switcheroo Mori."

The Netherlands offers Lucy Quirina de Berk. In 2002, she was initially suspected of having murdered thirteen patients by drug overdoses. Then forty-one years old, the nurse was officially charged with eighteen counts of murder and attempted murder at four hospitals in The Hague. In March 2003, she was sentenced to life in prison after a panel of judges found sufficient evidence of her having killed three children and one elderly woman. However, concerns were raised as to the validity of a toxicology report and whether the original trial was unduly biased. In 2008, this led to her release from prison, and in 2009 she began a new trial before the Arnhem appeals court.

Prosecutors said de Berk was obsessed with death and attempted suicide seven times over the course of the last decade. "[She] went about her work in a refined and calculating way when the chance of discovery was small. Apparently she believed she was qualified to hold the power of life and death over these people," said presiding judge Jeanne Kalk.

Germany is our next stop, where in November 2006, Stephan Letter, a young German nurse, was convicted of injecting twenty-eight patients with a cocktail of lethal drugs. The media nicknamed him the "Sonthofen Nurse of Death" after the quiet Alpine town where he resided. Letter is Germany's worst serial killer since World War II, and he received a life sentence. Letter claimed to have been motivated out of compassion for his elderly patients, but he was found guilty on twelve counts of murder, fifteen of manslaughter, and one of "mercy killing." The judge has determined that he "was at best superficially interested in the health of his patients." The state prosecutor has succinctly stated, "He killed as if it were an assembly line." Like Kristen Gilbert and Vicky Dawn Jackson, Stephan Letter came under suspicion because medications were missing at the hospital and an unusually high mortality rate was noted to occur on his hospital shifts.

It is unclear whether these murderers are becoming more common or merely more obvious. According to *USA Today*, cases of medical serial killers were almost unheard of until the 1970s, when four incidents were reported. The number jumped to a dozen in the 1980s and fourteen in the following decade. A 1990 book called *Nurses Who Kill* cites twenty-four nurses and nurse's aides charged with serial murder. Additional cases are continuing to be reported.

Of course, nurses are not the only medical personnel convicted as killers. Michael Swango, a physician, probably tops the American list. The handsome, debonair, blond-haired, blue-eyed, and extremely smooth-talking Swango was first convicted in 1985, after he mixed arsenic into the food and drink of an ambulance crew. All of the poisoning victims recovered, and he was incarcerated for only two years. Following his release from prison, Swango somehow managed to resume the practice of medicine.

In 2000, a federal court in New York sentenced Swango to life imprisonment for murdering three patients at a Long Island hospital. He is suspected of having murdered as many as sixty individuals over a twenty-year period. When asked for a motive, the principal prosecutor, assistant U.S. attorney Gary Brown, said, "Basically, Dr. Swango liked to kill people."

We can only visit two empty British prison cells when it comes to the most infamous of the international cases involving physicians. Back in 1957, Dr. John Bodkin Adams, a general practitioner from Eastbourne, England, admitted during a murder trial to "easing the passing" of some of the old ladies who died under his care. Adams is suspected of having killed 160 individuals, and an investigation found that he was mentioned in 132 of his wealthy patients' wills, several times receiving bequests of Rolls-Royce automobiles—a fact that the English media found especially piquant. Oddly, it was in Adams' court case that the doctrine of double effect was first enunciated in British jurisprudence; Lord Justice Devlin addressed the jury and explained that a physician "is entitled to do all that is proper and necessary to relieve pain and suffering, even if the measures he takes may incidentally shorten life." Astonishingly, Adams was acquitted of the murder charges, but he was subsequently found guilty of thirteen offenses of prescription fraud, lying on cremation forms, obstructing a police search, and failing to keep a register of dangerous drugs. He died in 1983 during a hunting expedition.

Although Dr. Harold Shipman is now recognized to have been Britain's most prolific serial killer, the circumstances surrounding his crimes have not been widely discussed in America. Shipman resembled a bearded version of Robin Williams, and for many years this man was a popular general practitioner, with his friendly demeanor making him a respected and well-liked figure in rural Hyde, England.

At the conclusion of Shipman's trial on January 31, 2000, the judge told him, "None of your victims realized that yours was not a healing touch. None of them knew that in truth you had brought her death, death which was disguised as the caring attention of a good doctor." Convicted of fifteen counts of murder and one of forging a patient's will, Harold Shipman was sentenced to life imprisonment.

The United Kingdom was agog over the case, and the public demanded a fuller explanation and reassurance that steps would be taken to prevent recurrences. An official Tribunal of Inquiry, headed by Dame Janet Smith, began a thorough investigation of the matter. Smith's inquiry determined that Shipman's usual method of killing entailed administration of a lethal dose of diamorphine—a type of opiate. Like several of the other medical murderers, Shipman was a drug addict; he had been previously convicted more than two decades earlier of falsely obtaining medications. At that time he underwent psychiatric and substance abuse treatment but did not reveal that he had already secretly killed his first patient.

The tribunal has determined that Shipman murdered at least

215 people. His victims range in age from forty-one to ninety-three years old, but they were mainly elderly women. Unlike Adams, Shipman apparently only killed for profit in the single case involving the forged will. Otherwise, his motives remain unknown. In 2004, Shipman committed suicide in his maximum-security jail cell.

The above record of medical sociopaths and their crime sprees is by no means complete. Beatrice Crofts Yorker, dean of the School of Nursing at California State University, summarized for me her international review of ninety criminal prosecutions of health care professionals. She found that nurses accounted for 86 percent of the sample. Fifty-four of these individuals were convicted and are suspected of being responsible for a total of 2,113 patient deaths.

Dr. Michael Welner, a psychiatrist from New York University Medical Center, provided an explanation for why nurses predominate among the medical serial killers when he commented in the *Seattle Times*, "Nurses are aware of which medicine to get, when the shift changes, when they can be alone with a patient. . . . There is no screaming thrashing victim. There's just a silent poisoning. There's not so much as a whimper. It's as anonymous as a person drifting off to sleep."

This review of serial medical killers highlights a number of common features. Many of the doctors and nurses had long-standing mental health problems, including substance abuse, multiple past suicide attempts, incidents involving psychopathic behavior, and psychiatric hospitalizations. While euthanasia or mercy killing is sometimes offered as a motive, it appears in most of the cases to be a bogus excuse, merely part of the defense strategy. Lastly, killing sprees can go undetected for many decades. It takes at least one skeptical person—usually another medical staff member like Olga—to trigger an investigation.

12

Army of God

In 1986, Dr. Carl Kjellstrand wrote a seminal article for the prestigious *New England Journal of Medicine* about withdrawal of dialysis that was based on his experience with a sample of patients from Minneapolis, Minnesota. The article prompted a flurry of letters from his peers, many of whom praised his courage for daring to write about a practice that had quietly become commonplace. Other letters attacked him for deviating from medicine's obligation to preserve life at any cost.

Carl is a good friend of mine, and he generously offered to describe his subsequent experience. He is a tall, lean, athletic man who, despite approaching retirement age, can hike or cross-country ski at a pace that would grind most people into the ground. Born and educated in Sweden, he is a natural raconteur with a prodigious memory for literary quotations, a passion for social justice, and an overwhelming curiosity about people. When I called and asked him about his article's initial reception, he immediately had a fit of laughter and insisted on e-mailing me the following response:

The New England Journal of Medicine *editor must have been a bit nervous about the article's implications, because he phoned me a few days before the publication date to impart some fatherly advice about how to behave with reporters. We bumpkins who live west of an imaginary line that runs about thirty miles from the Atlantic Ocean's high-water mark are not to be trusted as having any judgment or culture.*

First, ABC-TV, or some such outfit, called from Chicago and wanted to come up and talk to a patient and me. This was easily arranged. A bunch of people then arrived with cameras and other stuff, and they peppered me with a lot of very good and sharp questions during an hour-long interview. They then descended on the patient.

To my amazement, the reporters said that the piece was going to be aired the same evening—they sure did not let the grass grow. I sped home prepared to impress my lovely wife, Kerstin, and be ready for the calls from Hollywood. We turned the TV on and it finally came, all ten seconds of it. . . .

Disappointingly without any preceding fanfare, an enthusiastic reporter with the usual array of even, bright incisors stated, "A team of Minneapolis physicians report that dialysis patients are so miserable they would rather die than be on dialysis." Cut to a sad-looking Swede who mumbles something about how people on dialysis can have a tough life and prefer death. . . . Cut to an excited patient exclaiming, "I don't want to die!!!!!!!!!"

Sometime later, the telephone rang, I picked it up, and a strange voice inquired, "Is this Dr Jellystrand?"

Despite the caller's obvious confusion about the name, with my accent it was hard to deny that he had reached the celebrated foreign nephrologist.

Having clarified my identity, he went on to say, "We have heard about the murder of your patients. The Army of God is prepared to carry out His will. We have a bullet with your name on it."

To which I responded (and I am still pretty pleased with myself),
"Thank you very much for alerting me. Just so you know, I was the
pride of the Swedish army and have several gold medals attesting to
my skill at killing people with mostly small-caliber guns. Make sure
your first shot is a good hit—because it will be your only chance.
Now go to hell, where you belong!"

> *Best as always.*
> *Carl*

When I telephoned Carl to thank him for the anecdote, he immediately quoted Winston Churchill, who is purported to have said during his time as a correspondent in the Boer War, "There is nothing more exhilarating than being shot at without effect." To this Carl added, "Or having someone threaten to shoot you."

We discussed a number of cases in which dialysis discontinuation ended up mired in the judicial system. According to a 1990 report citing data from the National Center of State Courts, there have been at least seven thousand court cases on termination of treatment, and at least fifty of those cases have been focused solely on the withdrawal of artificial nutrition and hydration from critically ill patients. However, Carl and I were aware of only three noteworthy cases involving dialysis cessation.

The first of these involved a seventy-seven-year-old man named Earle N. Spring. Spring had end-stage renal disease requiring five-hour hemodialysis treatments three times a week, and he also suffered from advanced senility. Heavy sedation was regularly required to overcome his resistance to treatment, and Spring often refused to be transported to the dialysis clinic from the nursing home where he resided. During treatments he would frequently pull the dialysis needle out of his arm, and his dialysis complications included leg cramps, headaches, and dizziness. One can imagine him at his dialysis clinic, disoriented and wildly thrashing about with blood flying around the room, or sometimes lying immobile in a drugged stupor on the dialysis couch.

In January 1979, his wife and only son petitioned the probate court in Massachusetts for an order directing that dialysis be terminated. The son was appointed as the temporary guardian, but, in view of the request to stop treatment, the court also appointed a guardian ad litem to represent the elder Spring. A guardian ad litem is a responsible individual—often a social worker or attorney—assigned by a judge to assist the court in determining the circumstances and provide independent advice. The probate court judge ruled in favor of treatment cessation, but this decision was appealed by the guardian ad litem. When the intermediate appellate court in Massachusetts affirmed the probate court's decision, the guardian ad litem appealed again. After fifteen months of hearings, appeals, reversals, and stays, the Massachusetts Supreme Judicial Court finally issued its decision affirming the lower courts' rulings—one month after Spring had already died.

During this torturous period, the newspapers were filled with bitter debate. Representatives of the nursing home made clear their worry that if Earle Spring was permitted to stop treatment, this would open the floodgates and all of their other demented residents would be permitted or encouraged to die. An interview with one of the nursing home aides resulted in a bold headline declaring, "Earle Spring Says: 'I Want to Live!'" An attention-seeking fringe politician entered the fray and proclaimed that on the day Earle Spring breathed his last breath, he would demand the district attorney indict the patient's nephrologist, Dr. Lee Shear, for murder.

Several years after the case was resolved, when Shear was on the cusp of retirement, I had an opportunity to interview him. The Supreme Judicial Court had declared in its final ruling: "Little need be said about criminal liability; there is precious little precedent and what there is suggest that the doctor will be protected if he acts on a good faith judgment that is not grievously unreasonable by medical standards." Despite these words, in the interview Lee appeared to me to be a haunted man. Visibly trembling, he glanced sideways around my office and said, "I sincerely wish that I had never gotten involved!"

The second legal case took place in New York and involved a patient named Peter Cinque, a forty-one-year-old man who, in addition to his renal failure, also suffered from multiple diabetic complications including blindness, partial amputations of the legs, a bleeding gastrointestinal ulcer, and cardiovascular disease. Much of the time he was in considerable pain, for which he required potent analgesic medications. Otherwise, he was clearheaded and mentally alert.

In October 1982, because of the pain and other symptoms, Cinque decided to stop dialysis and die at home in the presence of his family. He discussed this decision with his relatives and with a Catholic priest. He had several conversations with the nephrologist and the Lydia Hall Hospital administrators, and each of them initially agreed to his request. As one of their conditions, Cinque temporarily stopped all of his sedatives and pain medications so that he could undergo a competency evaluation by two psychiatrists. They both concurred that he had the mental capacity to make the clinical decision to end treatment. On that same day, Peter Cinque signed legal documents further articulating these wishes.

His brother, Mark Cinque, commented that Peter was comforted and relieved to know his preferences were to be respected. However, several hours later, the hospital not only refused to terminate the treatment but also went to court to request a continuance. Mark Cinque described his brother as then going through "the most painful period of agony and suffering I had ever seen."

Two days later, Cinque went into respiratory arrest, lapsed into a coma, and was placed on a respirator. The guardian ad litem appointed by the court agreed with the family that Cinque had left unequivocal expressions of what he wanted prior to becoming comatose. The lower court found in favor of stopping the ventilatory and dialysis treatments, but the case was immediately appealed. The New York Court of Appeals, the highest court in the state, finally concluded that Cinque's wishes not only were "clear and convincing" but also met the more stringent standard of being "beyond a reasonable doubt." The hospital was ordered to stop all life support treatments.

In a cruel turn of events, the hospital acted posthaste upon learning of the court's ruling, without allowing the Cinque family to be present at the bedside. He died without having any of his loved ones in attendance.

The Peter Cinque case led Willard Gaylin, president of the Hastings Center (America's most prominent bioethics organization), to inquire, "How did we get into this 'Alice-in-Wonderland' world, where a man must beg for his legal rights, prove his sanity, endure court hearings and finally be reduced to a living cadaver to do what has generally been accepted as his privilege?"

The third legal case took place in Hennepin County, Minnesota, where Carl Kjellstrand practiced for a number of years. According to Carl, "In the early 1970s, a colleague of mine was treating a sixty-two-year-old woman—whose name I can't recall—with peritoneal dialysis." An alternative treatment to hemodialysis, peritoneal dialysis involves repeatedly placing and removing special fluids into the abdominal cavity, rather than hooking up the patient's circulatory system to a machine for blood cleansing. A plastic tube called a catheter is surgically inserted into the abdomen to allow for the process.

"Like Earle Spring," Carl explained, "she had become progressively demented and unable to care for herself at home. In the hospital she would scream for no apparent reason, and her behavior was increasingly disturbing to the other patients. On numerous occasions, she pulled out her peritoneal dialysis catheter. Much of the time, the woman had to be restrained to prevent this from recurring, since dialysis was not possible without the catheter. After many well-documented discussions with her husband, the doctor and the spouse agreed to discontinue the dialysis. She died shortly thereafter, and six months later her husband also died.

"Three years following the patient's death, two of her children accused the nephrologist of murder, but were unable to convince authorities to pursue criminal charges. The children then filed a civil lawsuit against the nephrologist and hospital, and argued that the physician had 'abandoned' his patient by stopping the dialysis. They sued for a million dollars; however, the jury took about five seconds to find the defendants were not liable. Furthermore, given the patient's poor health, disability, and dire prognosis, the court declared that even if damages had been awarded to the plaintiffs, there would have been no monetary value—zero dollars and zero cents—ascribed to her death. The doctor was clearly not guilty of murder or of malpractice."

In talking to Carl about these cases, there were several take-home lessons about what transpired. The realistic risk of liability is minuscule when decisions to terminate treatment have been well documented and managed in a clinically appropriate fashion. But the unfortunate truth is that medical staff will never be granted complete immunity from civil or criminal liability. When it comes to stopping dialysis or curtailing any other life-prolonging treatment, there will always be a risk of accusations, investigations, and penalties. Not to mention that, as Carl pointed out, there are also some wackos out there from the Army of God.

Dr. Robert Truog, director of clinical ethics at Harvard Medical School and a physician at Children's Hospital in Boston, has come to a controversial conclusion that resonates for me. In a medical journal article entitled "Are There Limits to Withdrawing Life Support?" Truog wrote:

> I believe the notion that we only allow patients to die and that we never kill them is incoherent and has led us to greater confusion when thinking about end-of-life care. Let me be clear, I would never say to a family that we are going to kill their loved one by removing the ventilator. My point is that the only language we have to describe what we are doing [is] the language of "allowing to die."

When Bobby Schindler stated that his sister was killed, I can sympathize with this position because of this linguistic confusion. Whether one says that death is accelerated or life foreshortened, the bottom line is that our actions lead to people dying sooner than they would otherwise. On an emotional level, it may seem that by accelerating death doctors and nurses are participating in killing. However, not for one moment would I suggest that this sort of killing has anything to do with murder—there is certainly no malice intended, and that is a key component of the definition. To suggest that withdrawal of a treatment is the equivalent of murder is a grievous mistake. But death-hastening medical practices almost always warrant a degree of unease.

I am a great admirer of Robert (Bo) A. Burt, the Alexander M. Bickel Professor of Law at Yale Law School. Bo has a long-standing interest in end-of-life medical practices, and he has vividly described an underlying aura of wrongdoing surrounding death. Beginning with the story of Adam and Eve, Bo cites a wide range of evidence to support his hypothesis that in our Western cultural tradition death is viewed not merely as a fearful event but one that is intrinsically shameful and even immoral. He has pointed out that humanity's first sin led to the inevitability of death—Adam and Eve's punishment for disobeying God's commandment to refrain from eating the fruit of the tree of knowledge was to be expelled from the Garden of Eden and to lose eternal life. Bo speculates that there is an intensification of guilt-ridden attitudes among physicians directly engaged in the dispensation or occurrence of death. I think he is correct. Even when I feel entirely justified and confident about the appropriateness of my own role, I also feel some degree of distress and guilt. And that may actually be a good thing.

Dr. Joanne Lynn is one of the most highly regarded authorities in the field of palliative medicine, and the principal investigator of its most influential research investigation, called SUPPORT. Among other things, this study documented the disconnection between how Americans want to die (at home with minimal fuss) and how their lives actually end (during hospitalizations in the midst of multiple interventions). At a conference in Madison, Wisconsin, I listened as Joanne told her attentive audience, "It is all right to feel uncomfortable about what we do. The clinical decisions we arrive at about our patients' lives should make us feel uncomfortable. They should never become routine or feel comfortable—because if they do, then we are in real trouble."

Were Amy and Kim too complacent when it came to the circumstances and decisions made during Rosemarie Doherty's final hospital stay? I genuinely do not think so. But, unfortunately for them, their case proved to be instructive in a lot of unintended and unexpected ways, eventually demonstrating that while the debate rages between the pages of high-profile medical journals and legal briefs, there is a real-world price to be paid for our inability to reach a consensus on these delicate issues.

13

Amy Goes Downtown

During the visit by the state police to Amy Gleason's house, the two detectives remained adamant about wanting to continue the interrogation at their office. Amy did not prefer to leave her comfortable home, but she was by nature inclined to be helpful. She also couldn't find a reason to refuse their demands.

"At that point I am scared to death," she said, "and naturally my irritable bowel condition flares up. So I tell them to hold on while I run to the bathroom, and they snort, 'We really need to go downtown.'

"And then I get sick again. So I am in the bathroom, they are in my kitchen, and I am thinking, 'Holy shit, what am I going to do now?' So I come back out, only to hear them repeat their mantra: 'We really need to go downtown.'

"I ask, 'I don't suppose I can take a bath first?' I know it is a ridiculous thing to say, and they immediately respond, 'No, you cannot take a bath. You need to come downtown.'

"I am practically in tears, and begin explaining, 'I just finished a twelve-hour shift at the hospital. I am completely exhausted and my supper is cooking. Can I eat something and then meet you later?' They are clearly not interested in a tale of woe.

"I suddenly realize that my husband will not understand what is going on, and I exclaim, 'My God, what am I going to tell my husband?' They suggest I write a note that says, 'I am at the state police.' I think that when my husband comes home and reads this note he is going to have a heart attack, but I quickly scribble that everything is fine and leave it on the counter for him.

"I go out the door and down my front steps with these two huge guys. Both are dressed in trench coats, and although they had shown me their badges originally, it is incredibly scary to be with hostile strangers who are walking to their unmarked police cruiser. They point at my car and tell me to follow them, and we head on Route 91 in the direction of Springfield and get off at State Street by the state police barracks.

"I pull into in an illegal space, because that is right behind where they park. I get out of my car and say, 'Oh, my goodness, I am in an illegal spot. Can I park here?'

"They shoot me a look and say, 'You are with us. One would think it is okay.' However, all I am thinking is, 'What do I know?' I flash on the thought that I am going to come out and discover my car is towed.

"I go into the building, and it is horrifying. It is eight-thirty at night, so there is nobody else around. It is dark, and we go in this rickety elevator up to the second or third floor. I begin to think, 'I am with these two humongously big men in a strange building. I have no idea what the hell is going on.'

"I sit in an uncomfortable chair and one guy is asking the questions and the other guy is pounding on a laptop computer, typing everything I say. Trust me, he never took typing classes. He does not know how to freaking type, and it is driving me crazy.

"Of course, I have to get up fourteen times to go to the bathroom. The first time I say, 'I have to go to the bathroom again,' it precipitates an argument.

"The policeman who is interrogating me retorts, 'You just went.'

"I say, 'It's coming out either here or there, so make your choice.'

"In between trips to the bathroom, I am sitting right in front of the guy who is typing—if you can call it that—and I ask him, 'Would you mind moving your gun?' His pistol is sticking out of a holster on his side and is staring me in the face. This also turns into a struggle.

"'Don't look at it,' he snorts.

"I say, 'It is really offensive.' Frankly, I cannot stand staring at this big, huge weapon, and I am scared to death being down there with these two guys. I explain, 'Can you just move your gun, because I will feel better if I don't have to look at it?'

"'No' is all he says, ending that discussion.

"Around then is the moment I begin wishing they would both come down with renal failure. To this day, I have not seen them in the hospital and they have not been assigned to me as patients, but I am still hoping. For that matter, I would not mind if the district attorney, Bill Bennett, also develops renal failure, because it was Bill Bennett sending them after Kim and me."

Listening to Kim Hoy tell it, the days leading up to Rosemarie Doherty's death were a flurry of activity.

"Rosie's son calls a couple of times, and I update him about how she is doing." Kim begins. "The family knows about her DNR status. I explain about the pain medication and how I had asked her doctors to change it. I am trying to keep her comfortable and to also take care of the other people on the unit. Rosie is not going to go on Saturday—not on my time. She hasn't reached that point. But you can tell she is getting there quickly.

"The day goes by trying to keep her bottom clean, positioning her, putting pillows under her arms, and talking to her. When things quiet down, I come in midway through the shift—and I love Kurt Vonnegut—so I kerplunk down alongside of the lady in the bed next to Rosie's and read both of them a short story. Sometimes at work I hide. If you come on the renal floor and can't find me, don't even look. I am behind one of those closed doors and am probably reading. I read to my patients."

Kim laughed mischievously and whispered, "Don't tell anybody." She giggled and then continued, "I read them a story by Kurt Vonnegut called 'A Long Walk to Forever.' It is soooo romantic! Girls love it! And they both loved it—Rosie and her roommate. I put the head of the bed up and say, 'Scoot over!' I get right down into the bed with her roommate, lean over the pillow, and begin reading the story. Amy may have come in for the last half of it. But it is just business as usual.

"Well, that was Saturday. I go home. On Sunday, I come into work and have the same patient assignments. I get the report about the night. Rosie has had some changes. She is starting to get a little mottled in the feet. She is making less noise. Things like that. So basically I concentrate on keeping her comfortable. The real issue this morning is her respiratory rate—she keeps going up into the thirties. Rosie gets short of breath, and you can just see the muscles between her ribs contracting and releasing—she is just working everything she has to breathe. That becomes a real focus for why I am medicating her with the morphine. It is less about the physical pain and more about dealing with the respiratory rate and trying to get those respiratory muscles to relax a little bit. It is pretty awful. I feel so badly for her.

"I also have to keep reminding the TA [Olga] with whom I am working not to put on the oxygen. I have to repeat, 'Please, do not put the oxygen on her.' I go into Rosie's room, and the oxygen is on and I take it off. I just cannot get Olga to understand about oxygen and emphysema."

Kim paused briefly, looked across the table at me, and said somewhat defensively, "After a while, however, I suggest we put on a shovel mask with humidification, because her respiratory rate is so fast. Rosie is mouth breathing, and she is getting all dry. I can't stand that. That has to be the worst thing you ever have—a dry mouth and you can't do anything about it. We do some humidification, and she seems more comfortable.

"Rosie passes that day. She passes on Sunday. Olga comes out and says, 'I think she is dead.' I walk into the room and check Rosie's pupils and listen for an apical heart rate. As I am coming out of the room, Dr. Hwang, the internist, arrives to check up on Rosie's condition. I say, 'Well, your assessment has just turned into a pronouncement.'"

Kim and I were sitting at a table in the back of the restaurant with a window facing a grove of trees, which were just beginning to sprout green leaf buds. It had been an unusually long winter in New England, and the sight was welcome. We were engrossed in the story and unaware until that moment there were nearby railroad tracks running through the trees. A train began to rumble loudly, and conversation halted until it passed.

When Kim finally resumed talking, she was preoccupied. Her voice got raspy as she spoke. "Earlier in the morning I'd spent a few minutes with Rosie, telling her that she was not alone. Her son Michael was going to come and see her later in the day. But I told her she did not have to wait for him. I said, 'If you do not want him to see you this way, you do not have to wait for him to come. He will understand. If you want to go now, you can go now, and I will be with you. You do not have to wait.'"

I was skeptical hearing this interchange. I realize it is not uncommon for hospice and other medical personnel accustomed to treating dying people to say such things. It is probably my own lack of experience, but I am not convinced individuals can control when they die. I suspect caretakers utter these words believing they are expressing them for the patient, but they are really an incantation that comforts staff. Then again, I do not know what it is like to spend the final hours with dying individuals.

Kim continued, "Olga overheard me saying this to her, and apparently she took offense. But I wanted Rosie to know if she was reticent to have her son see her that way, it was all right. He would be okay with it. If she wanted to wait for him, then he would be okay with that, too.

"Usually I wipe patients' foreheads a little bit. I talk to them quietly. Sometimes I say a little prayer with them. It depends on the person, and it depends on the situation. I kind of play it by ear. On this final day, Rosie was not talking, but I think she knew what I was saying."

Kim was quiet for a moment. "Well, Rosie passes, and Dr. Hwang does the pronouncement. And we go in to do postcare. I know the newfangled way is to just put the patient in the bag. Period! But I still have to wash them. Until the day I die, I will take care of them. I will wash them.

"I have my own little things. Usually I turn on a radio station. I tell the patient what I am doing as I do it. I just talk to them. And I always, always, always do the head transfer. Always! I never want anybody to slip. I tell staff you have to be careful when you move patients onto the stretcher. The morgue's stretcher is metal, and it infuriates me if people are not careful. I make sure that I get the head, because I would have to kill somebody if they were careless and a patient's poor little head smacked down on there."

A gaggle of waiters and waitresses materialized a few tables away, and suddenly they were joyfully singing "Happy Birthday" to another diner. Kim ignored the singing and emphatically repeated to me, "I have to hold the head, because I would be extremely upset and annoyed beyond belief if I ever heard a head slide over and—" She made a noisy *dulllk* sound. "I would simply freak out. You just do not do that. Sometimes staff forgets it is a person. That"—she points to an imaginary body on our table—"is a person. I remind them!"

Kim explained that she always requires an assistant when she does the postcare, because it is difficult to maneuver a body by oneself. She went on to say, "When I take a line out from the intravenous and other tubes, I put pressure on the puncture site. The puncture site will sometimes ooze. Afterward, I put on a little gauze dressing. Most staff handle this differently." She imitates the whooshing noise of someone quickly and carelessly pulling a tube. "I cannot do it that way. I am also unable to zip the bag. I can never be the one to zip it over the face. Whoever is helping me has to do that job. I just can't do it. I just can't.

"I finish Rosie's postcare. Olga and I cry. Every time it happens, I tell myself that I am not going to cry. I cry every time. I am such a schmuck."

Kim sat back in her seat and aimlessly pushed a shrimp around the plate. "I ask Olga to zip up the last part of the bag and she does it." Kim pulled an imaginary zipper up her neck and over her head while making a loud zipping noise. "Then we call the orderly to come and bring Rosie down to the morgue."

Rosemarie Doherty's hospital chart is about ten inches thick. It offers another window on the end of her life. The medical chart is, of course, intended to be a means of communicating with colleagues and a record of the medical situation. When I was learning how to write notes in hospital charts, I had a singularly brilliant and paranoid supervisor named Dr. Lew Glickman. Lew pounded into me his belief that every word should be written as if the patient were reading the chart over your right shoulder and his or her lawyer were reading it over your left shoulder. In his hospital progress notes, he quoted his patients at great length in order to avoid any possible ambiguous interpretations.

Medical charts are not generally written this way, but instead are succinctly filled with facts and devoid of opinions or details that might capture the uniqueness of people. For instance, by reading a chart you will never know if a male patient's finest moment came when he served in Korea, or whether a female patient gave birth and devoted her life to raising a developmentally delayed daughter, or whether those jailhouse tattoos were received while serving time for armed robbery. Not surprisingly, hospital records today provide a thoroughly desiccated and jargon-filled description of patient illnesses and complications.

Rosemarie Doherty, a sixty-seven-year-old woman, was admitted for the last time to Baystate Medical Center for what turned out to be a six-week hospitalization. She arrived in an ambulance after developing a gastrointestinal bleed. During the previous month and a half, she had been residing at a rehabilitation facility recuperating from a motor vehicle accident in which she had sustained multiple broken bones, including her pelvis, ribs, and sternum. In between, she was briefly hospitalized to repair a clotted fistula—a special tract that is made in the arm of patients by connecting an artery and a vein to provide access for the large needle inserted to perform hemodialysis. She subsequently had bleeding in the back of her abdomen that resulted in a partial blockage of circulation. The final hospitalization was complicated by an infection with the bacteria C-Diff (Kim was correct about the cause of the diarrhea) following a course of an antibiotic for a skin infection. She also had a bowel obstruction from her compromised vascular system.

Because of long-standing emphysema and congestive heart failure, she had continuing respiratory difficulties. An endoscopy showed a blood clot in her stomach but no obvious cause of bleeding. Colonoscopy confirmed the obstruction and infection with C-Diff. Despite comprehensive medical treatment, she gradually declined. She was intermittently confused and disoriented, and she had persistently bloody diarrhea. After an exacerbation of the congestive heart failure and further difficulties from the intestinal obstruction, the staff and the family agreed to discontinue hemodialysis and to more vigorously treat her with morphine.

Rosemarie Doherty died two days later. The final note from the intern breaks with the usual medical jargon by concluding, "May she rest in peace."

In the last few pages of the hospital chart is a petition, entitled *Commonwealth v. John Doe*. It states that a copy of the medical record is officially requested by the district attorney's office for the superior court grand jury to be held on December 20. As grounds for this motion, the Commonwealth explains, "Charges relating to murder . . . are currently pending in the Superior Court."

14

More Murder Accusations

Until I heard from Amy, Kim, and Olga, I had no inkling that
health care professionals could conceivably accuse each other
of committing murder, let alone be able to enlist law enforcement in
determining guilt or innocence. Nurses and physicians do not anticipate a visit by the police. Certainly no one seriously considers that
he or she may incur criminal penalties. However, during the past
several years I have met a number of medical staff from around the
country who not only were accused and investigated but lost their
jobs and in some instances were sent to prison before finally being
found innocent of the charges. In every instance their lives were ir-
reparably impacted by these controversial disagreements.

Sharon LaDuke is one of these accused medical professionals. I went
to visit her in the small town of Bountiful, Utah. LaDuke accurately
described herself as being "a curly-haired grandmother," and her
blond locks were distinctly tinged with gray. During our meeting,
she looked exhausted from having just flown back from a consulting
job in North Carolina. Despite the fatigue, Sharon freely admitted to

being the kind of person who would not hesitate to speak her mind, a character trait that got her into major trouble.

The short version of the story is that LaDuke was the nursing director of a tiny, rural intensive care unit in the city of Ogdensburg, New York, near the Canadian border. After administering medications to a dying patient who had been withdrawn from a ventilator—as ordered by an attending physician—a nursing administrator formally accused her of euthanasia. The hospital reported the incident to the state department of health, and the matter was referred to the state board of nursing and the county district attorney. After an exhaustive investigation, the district attorney saw no grounds to indict her for murder, but she was still fired from the hospital. She was vindicated only after a prolonged and expensive civil lawsuit.

Immediately following the patient's death, Sharon agonized over whether she had inadvertently committed euthanasia. In an eloquent essay published in a medical journal, Sharon described how immediately after the patient's death she ruminated over the *E*-word:

> When Willie's suffering was over . . . I was alone with my thoughts and began to question myself. It is widely known and well documented that nurses and physicians can feel guilty after ordering or administering analgesics and sedatives to people who are dying. That's because these medications have a "double effect": As they ease or end the symptoms associated with dying, they also can potentially cause vital signs to deteriorate—in essence, hastening death. Many clinicians have trouble on a moral level distinguishing between administering medications that might hasten death—an act that is required if the dying are to receive appropriate care—and giving drugs in order to hasten death, which is euthanasia.

Five years after the death, Sharon told me, "It will probably always be the single most important event of my life, but I'm glad it's far behind and that my friends and family stood by me."

The longer version of this tale was recounted by Frank Dobisky, the son of LaDuke's purported victim. Frank is an unassuming middle-aged man with glasses and thinning brown hair. Frank's family is among the oldest and most respected in Ogdensburg, which was a finalist for the 2006 National Civic League's All-America City Award. Ogdensburg's community center is named after Frank's father, who aside from owning and operating the local department store also served in a number of civic roles, including the volunteer rescue squad and the chamber of commerce. Frank's mother, Willis Dobisky, was a stalwart member of the Presbyterian Church and sang in the choir. She also participated in the hospital auxiliary, helped out in the gift shop, and gave birth to four children at this same medical facility. Mrs. Dobisky was seventy-two years old at her last admission.

According to Frank, the hospital administrator who accused Sharon of euthanizing Mrs. Dobisky "picked the wrong people to fight." He explained "that on the final night of my mother's life, she had been in the hospital for a week or ten days and was on life support. She was put on life support even though she had a health proxy asking that it not be done—but the person who held the proxy was not around when she came into the emergency room, and when they asked my mother if she wanted to receive the treatment she was in pretty rough shape and said yes. So they connected her to a ventilator.

"She was not improving. On several occasions she had told our whole family that no matter what happened she did not want to live on a machine. We met with the doctor and said, 'So, what are her prospects if we take her off life support?' The doctor explained that it would probably be a matter of hours before she would just slip away."

Frank and his family agreed to take her off the ventilator, but on condition that the doctor promise to keep her comfortable and ensure that she did not experience any pain or discomfort. It seemed like a straightforward solution, but the reality proved more difficult to orchestrate.

"They took her off the ventilator," he said, "and she was having trouble breathing and was obviously very uncomfortable. We asked the nurses during the night to increase her pain medication, and they replied, 'Well you know if we do it, we'll kill her. We can't do that.'"

Given the assurances from the doctor about managing Willis' symptoms, the family was not expecting to hear that unsatisfactory answer from the nursing staff. By the time Sharon came on duty the next morning, the family was frantic. Sharon was the nurse in charge of the ICU and the minute she arrived the family approached her.

"She looked into it and talked to the doctor, and the doctor told her to increase the pain medication and make sure that Mom was okay—which she did," Frank recalled. "The doctor, however, did not sign the order. In any event, Sharon stayed with the family and she made sure that my mother was comfortable. Early that afternoon my mother peacefully slipped away. And when she died, Sharon had all of us hold hands around my mother's bed and sing 'Amazing Grace.'"

At this point in his narrative, Frank's voice broke. He said to me, "And this is the woman that these fucking doctors and hospital administrators said killed my mother! Sons of bitches!" Crying softly, Frank apologized for his choice of language.

"In any event, Sharon took good care of us . . . she took good care of my mother. She was in the choir of my mother's church and had a gorgeous voice. Sharon sang at my mother's funeral.

"However, apparently Sharon was feeling kind of uneasy about what had happened and mentioned something to a colleague. A hospital administrator saw an opportunity to nail her, found out there was no doctor's signature on this thing, and suspended Sharon. . . .

[The administrator] eventually fired her, and then tried to get two different district attorneys to indict Sharon and us for conspiracy to murder.

"The district attorneys took a look at this thing and said, 'What was the motive?' They saw there was no money or estate involved. My mother had nothing when she died—just Social Security. We certainly did not want her to die. Her health was such that she was failing and she had been ill for some time.

"The hospital took a hard line. Sharon tried to talk to them and she was getting nowhere. We asked the Joint Commission to look into the matter." The Joint Commission is an independent, not-for-profit organization that accredits and certifies health care organizations and programs in the United States. The commission was doing a routine inspection at the hospital when Frank Dobisky approached them.

Frank continued, "The family attended a meeting with the visiting accreditation team . . . who listened to us. One of the accreditors later caught us outside and was in tears over what happened, but no action was taken against the hospital.

"Nothing about this had been in the papers. We decided to let the hospital know we supported the nurse. We did not believe that Sharon had committed euthanasia on my mother. We filed suit against the hospital. We said that we were filing because it was unfair what they had done to Sharon and for them to put our family through this pain and anguish.

"The case dragged on. There were no negotiations. Since we were not getting anywhere in the legal system, I decided to take our case to the court of public opinion. The editor of the *Ogdensburg Journal* [the daily newspaper] and the *Advance News* [the Sunday paper] supported our position. The community was divided. This slowly bumped along until there was a change in the hospital administration. They came to an agreement with Sharon, who ended her suit and was then reinstated. The only reason we had sued was to support Sharon, and we dropped our civil action. I was glad that she came back to the hospital. I was not surprised when a few years later she left that place, and I was pleased she could hold her head high. She had nothing to be ashamed about. Sharon was God-sent. She is a terrific lady."

Dr. Robert Weitzel is a psychiatrist who had the misfortune—which is way too soft a word—to be accused of killing terminally ill patients. Weitzel was eager to talk and had been trying to recover not only from being charged with multiple counts of first-degree murder but also from having been subsequently convicted of manslaughter and negligent homicide. He was given a sentence of one to fifteen years in prison, and though the sentence was finally overruled, he had already served six months and a day by that time. He then spent an additional four months in a federal correctional facility after accepting a plea deal for charges related to poorly documenting the disposal of narcotics—wastage, the same problem faced by Kim Hoy.

I interviewed Robert several times. When I first spoke to him, it was four years after the nightmare had begun. He was sitting in front of his wife's computer, unemployed and waiting to hear whether his medical license would be restored. Even if he could once again legally practice medicine, he faced the hurdles of insurance companies that were unlikely to accept him as a provider, and medical centers and clinics that would be reticent to take a chance on employing him and potentially having to face adverse publicity. All of his assets had had to be sold in order to cover legal expenses, and he calculated that the trials cost not only his professional reputation but well over a million dollars. He was profiled in a March 2002 broadcast of *60 Minutes* entitled "A Sad Fact of Life." Weitzel was trying to write a book and had created an elaborate website detailing the ordeal. The site included detailed medical records and transcripts from his trials, amicus briefs, letters of support, and a podcast and script of the television broadcast. The website concluded with a quotation from Albert Schweitzer: "Pain is a more terrible lord of mankind than even death itself."

Robert recounted the following to me: "My case began with wastage. I am a psychiatrist who came from Texas and set up a practice in Utah. One of my jobs was in a newly organized headache clinic. Admittedly, it did not have an official log of medication wastage—a federal requirement. When there was some leftover narcotic, we would document the action in the patient's chart—but that is still not the same thing as a log of the wastage. Consequently, when this came to light, I was technically guilty of 'obtaining controlled substances by deception,' and was then subjected to intense scrutiny by the Drug Enforcement Administration, the state licensing bureau, and the Justice Department. Inquiries were subsequently made at the hospital where I was the associate medical director of a geropsychiatric unit, and they turned up a disgruntled nurse who was eager to accuse me of purported misdeeds.

"Back in 1995–96, we had a run of demented folk from several nursing homes referred to our rural hospital in northern Utah for assistance with agitation and confusion. Some of these patients developed acute medical illnesses. In several cases, I had discussions with family members and decisions to defer curative attempts at treatment and to instead institute palliative care. Often these patients were moribund and unable to interact coherently, and if they were in pain they received modest doses of morphine. I am not an advocate of euthanasia, and what I am talking about are five-milligram doses of morphine, not enough to kill someone—in fact, hardly enough to kill a mouse. Certainly none of the patients were overdosed, and none died immediately after receiving pain medications. However, by the time the investigation was completed, I was charged with five counts of homicide.

"The press demonized me as being a mass murderer who killed helpless old grannies. No one was interested in hearing the truth. If convicted of first-degree murder, I faced life imprisonment. I am too polite to tell you my unvarnished opinion of the jury system, but in their dubious wisdom they found me guilty of manslaughter and I was remanded to prison."

Dr. Perry Fine is a pain specialist who was asked early on by the district attorney's office to serve as a consultant and review Robert Weitzel's case files. Perry concluded that reasonable palliative care was provided for each patient, and he recommended that the state of Utah not pursue criminal charges. Not only was his recommendation disregarded, but it was also not made available to Weitzel's defense team. Perry told me that during the sentencing stage of the trial, he was away on vacation. It was only upon returning and serendipitously attending an ethics seminar held at a colleague's home that he mentioned his role in the investigation—and then heard the next morning from a very surprised and irate defense attorney. After Perry's consultation became public knowledge, a judge ruled that prosecutors had failed to disclose powerful and credible evidence. Because the report had been buried, Robert Weitzel's conviction was overturned; he was released from prison and granted a new trial. And although prosecutors offered to reduce charges against him to five misdemeanor counts of negligent homicide, Robert refused that offer, saying he'd rather face them in court. In the second trial, he was found to be innocent of all charges.

The subsequent *60 Minutes* broadcast begins with Robert polishing glasses and setting tables in his temporary job as a waiter. It concludes with Ed Bradley turning to Perry Fine and asking one final question: "In your opinion, what's the importance of this case?"

Perry replies, "An individual has been charged with murder who in his best estimation was practicing a legitimate form of medicine.

And if that can happen to him, then it can happen to any physician. This is absolutely the wrong message to give physicians. They will pull back. And who will suffer is you, me, and anybody we care about."

During a visit to Salt Lake City, I had an opportunity to interview an assistant attorney general, Charlene Malone (who communicated purely as a private citizen and not as a spokesperson for her office). Malone remains convinced to this day that the prosecution targeted a rogue doctor. She and her colleagues gathered information about Robert and determined he was unapologetically arrogant, lived a socially active lifestyle (certainly by Utah's predominantly Mormon standards), ordered elderly demented people to receive intramuscular injections of medications (this may be upsetting, but it is still standard practice in almost every psychiatric unit), and charged substantial (although ordinary) fees for visits with patients who were confused or comatose and minimally able to converse. The prosecutors concluded that he was a terrible physician, and they agreed that they would never have wanted Weitzel to treat any of their own loved ones. While obviously not grounds for murder, it is not surprising that investigators would vigorously (and in this case overaggressively) pursue an individual who appeared to them in a variety of ways to be unsympathetic and cruel.

Since the Weitzel case, the district attorney's office in Utah has not prosecuted any other physicians, and I am left with the distinct impression that after the spate of bad publicity, it is unlikely to do so in the immediate future. They recognize that the criteria for finding medical personnel guilty of criminal behavior are far stricter than those required in civil or malpractice proceedings. Put simply, cases involving deaths following the withholding or withdrawal of medical treatment or a "double effect" of medications are fraught with complexities that may resist being held to the beyond-a-reasonable-doubt standard of criminal trials. Utah district attorneys are now compelled by law to seek the advice of a medical expert before proceeding to trial. Following the Weitzel case, the state medical society issued the following statement:

> The Utah Medical Association opposes the criminalization of medical care and sees unfounded accusations of physicians in criminal court and the criminal trial of physicians' professional judgment and quality of practice as a serious threat to patient care in the State of Utah and an unreasonable burden on the medical profession. Although it is acknowledged that the public must be defended against criminal actions, we do not believe that the professional assessment of medical competence necessary to discriminate between medical incompetence and criminal negligence can be judged fairly and knowledgeably before a lay jury in criminal court in the manner contemplated in *State v. Warden*. . . .
>
> Lastly, we believe that when a medical expert admonishes a prosecutor against filing a criminal complaint, it behooves the prosecutor to reconsider his position and seek the opinion of the Utah Medical Association, the Physicians Licensing Board, or some other regularly established and constituted panel of medical peers. Neither Utah's physicians nor their patients can afford this type of judicial embarrassment. It is a serious threat to good patient care for all Utah's citizens.

Meanwhile, Robert still struggles to move on with his life. When I reconnected with him a year later, Robert had moved to another state, where he was still without a medical license and was now employed as an office manager for a sympathetic physician. He was in the process of discussing with insurance companies the need to create and make available policies for doctors that cover the costs not only of malpractice but also of criminal proceedings arising inadvertently from medical care. He hoped that a new career could be created out of the debacle of his former profession.

One year after that, I received a desperate e-mail from Robert. The insurance idea had not panned out. He had then found a position selling electronic records software for emergency rooms, but had just been fired. A physician from Utah apparently had heard about his new job and called the company demanding that they take action against the "convicted murderer." Although Robert had originally provided the company with the entire history of his legal travails, this counted for nothing when they realized he could become a public relations liability. He had been hopeful the work would get his wife and him out of debt and on the road to building enough of a financial cushion that he could take time off to study for the test that must be passed in order to regain medical licensure. Whether you describe it as a Catch-22 or a particularly awful type of limbo, more than a decade had passed since the patient deaths that prompted the accusations against him, and his past entanglement with murder accusations still continued to haunt him.

Dr. Perry Fine, who figured in Robert Weitzel's proceedings and the *60 Minutes* broadcast, is a professor at the University of Utah's medical school, an anesthesiologist with advanced fellowship training in pain management, a former medical director of a local hospice in Salt Lake City, and currently the medical director and board member of the Vista Care hospice company. In addition to Weitzel's case, he similarly testified in early 2003 as an expert defense witness in *State of New Hampshire vs. John Bardgett*.

John Bardgett was a twenty-five-year-old nurse who administered morphine, Lasix (a diuretic), and other medications to two elderly, terminally ill women who were residing in a nursing home. He was then accused of murder by a coworker, a licensed practical nurse. She alerted authorities and even arranged for a phone tap to allow state police investigators to record her conversation with the unwitting Bardgett.

According to Perry, "In my opinion, John Bardgett was avidly treating the patients. [He was] providing palliative interventions to relieve the burden of dying and to prevent excessive suffering. The first case involved a woman who had end-stage liver cancer and required escalating doses of analgesics. The only obvious fault was that the nurse administered the medication intravenously instead of subcutaneously. But even the state's expert said he would have treated the patient the same way. The second case involved a woman who had fulminant [progressive] cardiac failure and severe dyspnea [shortness of breath] from pulmonary edema [fluid in the lungs]. Bardgett followed the doctor's orders, but despite treatment, the patient died an hour later. In February 2003, a Hillsborough County Superior Court jury deliberated for eight days and acquitted him of eight charges, including two counts of second-degree murder. They were deadlocked on four other charges."

In November 2003, as part of a plea agreement, Bardgett pleaded guilty to administering morphine without a doctor's orders. He was sentenced to two consecutive twelve-month jail terms with all time suspended, and he relinquished his nursing license.

I asked Perry for his opinion as to why some cases are litigated, much to the detriment of health care professionals, while others go uncontested. He had given this matter considerable thought and said, "The common denominator in cases that go to court is a practitioner who is not fully integrated into the community." He then referred to *The Scarlet Letter*, the Nathaniel Hawthorne novel about an adulterous woman singled out for public humiliation.

"John Bardgett," he told me, "is a man in the predominantly female profession of nursing. He has a tendency to shoot his mouth off. John Bardgett was a medic in the military and is a small guy who likes to be in charge, and he comes across as being cocky and arrogant. I understand he is perceived to be the sort of person who doesn't hesitate to inform others that he is correct and they are wrong. According to the local news media, when he was an emergency medical technician there were so many encounters with critical situations that he nicknamed himself the 'Angel of Death' [the same sobriquet accorded to Kristen Gilbert]. The newspapers had a field day with that!

"Similarly, Robert Weitzel was a physician from Texas who also never quite fit into the medical establishment of Utah. He arrived in a new community, dated a lot of women, and was seen as being a party guy. In his medical practice he sometimes walked on the edge—for example, treating headaches with opioids [narcotics]. Guys like Bardgett and Weitzel are outsiders."

Perry Fine's "outsider" theory is compelling and certainly seems to fit his two cases, but it is not supported by the story of Amy and Kim, or for that matter that of Sharon LaDuke. Each of these women was well liked in her hospital, and each was highly respected for being unafraid to advocate on behalf of patients. Sharon, for example, was clearly beloved in her small town—the local newspaper editorials fully supported her position, and the bereaved family even sued the medical center as a means of providing leverage for her reinstatement. While there is no doubt that some accused medical staff may closely meet Fine's profile of being outside the local norm or occasionally relying on maverick techniques, ultimately this hypothesis teeters on unfairly blaming the victim and does little to explain how Amy and Kim found themselves embroiled in a similar situation.

Having looked at a broader range of cases, I would instead argue that accused medical practitioners are simply unlucky. Absolutely anyone who treats the terminally ill could find themselves the object of allegations and legal proceedings, and there is an unfortunate confluence of factors that results in accusations. In investigating this issue I've found that the ranks of the accused include novices in the care of the dying as well as highly regarded leaders of the palliative care movement. Accusations can take place in small, isolated hamlets in the Midwest as well as world-renowned urban teaching hospitals. The complainant may be a principled individual or an unhappy malcontent. The district attorney may or may not personally adhere to strict moral or religious beliefs. Some law enforcement officials may be looking to prosecute and thus may welcome a potentially high-profile case. The accusation may occur at the same time the community is dealing with a disaster, such as Hurricane Katrina, affecting the provision of care to the very ill, or it can be sparked by the recent discovery of a health care professional who actually commits criminal acts, such as Kristen Gilbert.

Once a complainant appears and the police or district attorney's office begins an investigation, it becomes exceedingly difficult for them to stop. Professional and political reputations are at stake, and cases often seem to effortlessly fall into place. It is not difficult to enlist witnesses, be they coworkers who get swept up by the investigation and have reasons to cut deals or family members who stand to win large sums of money in subsequent civil lawsuits. Of considerable importance (not to mention concern) is the reality that once the process begins, virtually no clinician can expect to have all of his or her actions scrutinized and then emerge entirely blame-free. Hence the stakes for the medical professionals involved are incredibly high, since, as Robert Weitzel can attest, these are events that ruin careers and eviscerate lives.

Part of the difficulty, of course, is that pretty much every practitioner engages in some incorrect behavior or has some misunderstanding of a disease process over the course of their career. And yet accountability, in either medical malpractice or criminal justice, largely serves to discourage the accused from ever admitting to having made a mistake. The system forces medical staff to take the stand and declare they are totally innocent or entirely guilty, even though, of course, neither is likely to be true. Tossed into this mix is the public's confusion about complicated value-based issues that are at the heart of every case where patients have died after analgesic medications were provided or life support withdrawn.

Dr. Mike Germain, a nephrologist and my longtime research collaborator, has formed his own impression as to why Kim and Amy were accused, even though they do not meet Perry Fine's definition of "outsiders." "Neither of these nurses is exactly what Perry Fine calls an 'outsider,' but they are genuine characters who don't hesitate to speak up and cut right through the bullshit. I think that this country is terrified of death; these fearless women are a lightning rod for our anxieties about dying. Kim and Amy dared to carry on unafraid at work—unlike most people who worry about making waves or being different—and that unfortunately was what led them to be targeted by this ridiculous investigation."

When Dr. Robert Weitzel was sitting in prison following his conviction, an amicus brief was filed on his behalf by the Association of American Physicians and Surgeons. Founded in 1943, this national organization is dedicated to preserving the practice of private medicine and consists of thousands of physicians from all specialties. Their legal brief in his case is worth quoting. It begins:

> The prosecution of a physician for alleviating pain in a patient, without overwhelming proof of wrongdoing, would establish a dreadful precedent. Many patients have pain when they are near death. Accordingly, patients frequently die while they are taking prescriptions for pain relief. In some cases, the pain prescriptions may even unintentionally hasten death. It is simply unavoidable, and no basis for prosecution.

The Association of American Physicians and Surgeons has compiled a list of more than a dozen physicians who have been indicted for allegedly harming patients by administering analgesic medication or performing related actions. The cases of some of these doctors have been tried; others are still working their way through the legal system. The inventory of accused individuals does not include any nursing or health care personnel who may have been investigated by hospital ethics committees, licensing boards, or other noncriminal bodies.

Dr. Lloyd Stanley Naramore deservedly has a prominent place on the list, as his case is one of the most egregious. A two-count complaint was filed against him by the office of the Kansas attorney general charging attempted murder of a patient, Ruth Leach, and premeditated first-degree murder of another patient, Chris Willt. Both counts arose out of actions Naramore took in August 1992, when he administered analgesic and other medications to Leach and withdrew life support from Willt.

According to Stan Naramore, Mrs. Ruth Leach was a seventy-eight-year-old woman who was hospitalized with advanced breast cancer that had metastasized to her lungs, bones, and brain. Stan sat down with Leach's family in the hospital chapel, explained to them his clinical assessment, and spelled out the pros and cons of treating her pain with opiates. With the family's agreement and support, Stan told me, he returned and administered some analgesics. When Leach's respirations slowed and became irregular, her son suddenly expressed concern she was being euthanized. His worries were magnified when a hospital administrator approached him and suggested that the physician was attempting to kill his mother with an overdose. The patient was transferred to another community hospital, where she died three days later.

Mr. Chris Willt was an eighty-one-year-old diabetic with a pacemaker and a history of high blood pressure, kidney disease, and liver disease who had refused to continue his heart medication. Several days later he was brought to the hospital after losing consciousness at a local convenience store. Emergency personnel noted the presence of an irregular heartbeat and abnormal respiration. Willt was unable to speak and was diagnosed as having had a severe stroke. For several hours Stan and several medical technicians manually ventilated him—a physically exhausting process—because the diminutive St. Francis medical facility lacked a respirator. Deciding that continued treatment was futile, Stan consulted with another local physician, who arrived at the same conclusion. Ventilatory efforts were stopped, and Willt died eight minutes later.

Stan maintains his decisions were based in part on lengthy discussions he had had with both patients long before their emergency admissions. Following his arrest, Stan was unable to pay the $500,000 bond, and he spent the next eighteen months in jail awaiting trial. Stan was initially represented by a public defender. Then a friend came forward and contributed her modest life savings toward his defense, and they were able to secure the services of a devoted but inexperienced attorney fresh out of law school. The jury returned two guilty verdicts: attempted murder on the first count, and intentional and malicious second-degree murder on the second count. Stan was subsequently sentenced to concurrent terms of five to twenty years. A chilling documentary about the case aired on the A&E cable network, in which a juror explained that he determined Stan was guilty because he did not like the doctor's personality and therefore did not want him to continue practicing medicine.

As the sentence was pronounced in court, Stan's eight-year-old daughter shrieked loudly, and he cannot recall anything further until returning to his cell. Dumbfounded to find himself in a maximum-security prison and unwilling to cause his family any additional harm, he told his wife to no longer visit him and that he had never loved her. Stan wept while describing to me how he quickly signed the divorce documents and tried to accept that he would likely spend the rest of his life in jail. Penniless, he was serving concurrent sentences with no hope of being pardoned.

Before his nightmare began, Stan had never run afoul of the law. He had welcomed the opportunity to work in an underserved community in rural Kansas, where the nearest hospital was two hours away. In emergency situations, he would load patients into his private plane and fly them to places with better resources. He relished his medical practice and admired his neighbors, but his wife and he were Methodists in a conservative Catholic town. They were never accepted as being anything other than strangers. According to the medical writer for the *Kansas City Star*, he "didn't quite follow the code of friendliness and modest behavior . . . [he] played the ponies. Drank and smoked. Drove his flashy red Lincoln too fast."

In 1998, the state court of appeals reviewed the case and reversed Naramore's two convictions. The court also took the highly unusual step of directing a verdict of acquittal. This means that even with appropriate jury instructions, there was to be no retrial; whether or not they wanted to, the State of Kansas could not continue to prosecute him for these charges.

The justices expressed indignation that Stan had been criminally convicted, and they wrote:

This case adds a prospect of criminal liability to complex issues of health care for critically and terminally ill patients. . . . The burden of proof to establish criminal guilt of a physician for acts arising out of providing medical treatment is higher than that necessary to find medical malpractice or to impose medical licensure discipline. . . .

All three amicus briefs acknowledge the appropriateness of criminal responsibility where a physician's actions are clearly reckless or purposefully homicidal. However, they note that if criminal responsibility can be assessed based solely on the opinions of a portion of the medical community which are strongly challenged by an opposing and authoritative medical consensus, we have criminalized malpractice, and even the possibility of malpractice. [Dr.

Naramore's] case is a very good example of this. With no direct evidence of criminal intent, it is highly disturbing that testimony by such an impressive array of apparently objective medical experts, who found the defendant's actions to be not only noncriminal, but medically appropriate, can be dismissed as "unbelievable" and not even capable of generating reasonable doubt.

A number of years later, Stan's ex-wife—whom he described to me as "a beautiful young girl from Wisconsin"—sat in a multiplex cinema with their adolescent daughter and watched Denzel Washington in *The Hurricane*. With anguished bravado the protagonist, a falsely imprisoned boxer, tells his wife during her first visit to the penitentiary that he does not love her and she should petition for a divorce. The scene proved to be a revelation to Stan's ex-wife, who burst into tears, turned to her daughter, and exclaimed, "Your father really did care for me!"

In the darkened movie theater, the daughter looked back at her weeping mother and answered, "Well, duh."

15

The *E*-Word
Around the World

Most lay people and certainly most law enforcement personnel believe that opioid medications routinely kill people. Several of the aforementioned deaths that led to accusations of murder
were attributed to the administration of these drugs. Although the
public may think of narcotic analgesic or heroin overdoses as being
deadly, they mainly put people to sleep or sedate them. Deaths rarely
ensue—especially when one considers that addicts inject these substances multiple times every day. The rock stars and celebrities who
die following opioid overdoses invariably have taken combinations
of additional drugs, and their deaths usually follow accidental vomiting and aspiration. Palliative care practitioners downplay the le-
thality of opioids.

Instead, palliative medicine clinicians typically think of opiates as
providing relief from pain and amelioration of respiratory distress,
and they point to research data suggesting that narcotics actually

increase survival. Practitioners emphasize that patients who are already taking morphine or its derivatives quickly develop a tolerance that offsets many of the negative effects of larger dosages, and in these cases it is especially farfetched to suggest that even massive amounts of opioids are deadly. As Dr. Rob Jonquière, the director of the Dutch Voluntary Euthanasia Society, told me, "Morphine is a very good painkiller, but probably the worst way to end life."

This point is important because it counters the commonly held image that euthanasia typically takes the form of someone injecting and killing another person with an opioid. Even putting aside the false belief that narcotic medications are dangerously lethal, most American lawyers and bioethicists would not consider the previously described cases or Kim and Amy's treatment of Rosemarie Doherty to be euthanasia. Today in the United States, the withdrawal of life-sustaining treatments (such as dialysis discontinuation) and the administration of analgesic medications (such as morphine) for symptom management are both legal and ethical. A number of other practices, such as the cessation of artificial nutritional support or the use of palliative sedation, are also generally accepted, but they are admittedly somewhat more controversial.

Palliative sedation (also called *terminal sedation*) is probably an unfamiliar term for most people, and it warrants a more detailed description. In palliative sedation (which is relatively rare), health care personnel use medications to induce a coma in patients whose pain, shortness of breath, nausea, and vomiting are so excruciating they cannot bear to be conscious. According to a well-regarded study, between 5 and 35 percent of hospice patients have intractable symptoms in the last week of life, and a small number of these individuals may end up receiving a continuous intravenous infusion of the medication midazolam until they eventually die (usually after several days). To avoid even the appearance of killing people with the midazolam, periodically the rate of drug being administered is reduced in order to awaken the person and see if he or she remains in agony.

This procedure is unquestionably legal in the United States, and there is a growing national consensus that sedation to the point of comfortable sleep is entirely permissible. It is *not* generally considered to be euthanasia, and two of the most highly regarded palliative medicine authorities inadvertently ignited a minor firestorm a few years ago when they innocently referred to it in a journal article as being "slow euthanasia." Most American ethicists are in agreement that it is simply another acceptable technique for providing end-of-life care. According to a new policy adopted at the annual meeting of the American Medical Association in June 2008, there is no evidence that it speeds up the dying process, and physicians are now *obligated* to offer palliative sedation when "symptoms cannot be diminished through all other means of palliation, including symptom-specific treatments." Unfortunately, this report was issued too late to prevent a prominent palliative medicine physician, Dr. Paul Rousseau, from being investigated.

Rousseau was the associate chief of staff for geriatrics and extended care and the director of palliative care at the VA Medical Center in Phoenix, Arizona. A six-foot-tall, slender trail runner, he has a closely trimmed full beard and long hair that is parted in the middle and reaches to his shoulders; by his own admission he looks like he "never left the 1960s." The Veterans Administration may operate medical institutions that are particularly susceptible to political pressures, suspicions, and allegations: Paul is hardly the only physician from that organization I have interviewed, but most VA physicians are hypersensitive to their hospitals' close link to the government and are hesitant to go on record with their experiences.

Paul's professional nightmare took the form of an investigation about offering palliative sedation. From February until December 2003, his clinical performance was formally examined by the inspector general's office. During that time he felt almost entirely unsupported and shunned. He turned for assistance to Dr. Timothy Quill, who had previously challenged the medical establishment by publishing a moving but highly provocative article about assisted dying. Tim spoke from personal experience when he urged Paul to remain calm, saying, "I wish there were magical words to help you get through this, but there are none. You just have to believe in your heart in what you are doing and take it a day at a time."

Paul found he could manage that only up to a point. In exasperation during a conference call with several members of the National Center for Clinical Ethics, he exclaimed, "You people may not entirely agree with what I am doing. . . . You can sit there now when you are healthy and everything is going well, but you don't really know what it is to suffer. Feel free to be critical. However, I can assure you that when your time comes you are going to want someone just like me at your bedside."

During his ordeal, Paul was told he faced criminal charges and the loss of his job, and also that an undersecretary for veterans' affairs was personally involved in the investigation. He recalled repeatedly being compared to Dr. Jack Kevorkian by peers—a characterization he strongly resented. "The connotation," he explained, "is that I am a killer. I am not killing and I am not committing euthanasia. I am relieving suffering. My intent is entirely to relieve suffering. To this end I go through a whole process: I talk to the nurses, I get informed consent, almost every patient has a psychiatric evaluation, and practically every patient has a consultation by the hospital ethics committee. Palliative sedation is not something that I take lightly."

After the investigatory process was concluded, Paul remained dismayed, frustrated, and emotionally drained. "Even now that I have been cleared of the euthanasia charges," he remarked, "I am continuously being scrutinized. After having practiced here for nineteen years, I realize that I must leave the VA; frankly, I don't want to leave Phoenix and the stimulating environment of this particular medical center." When I followed up and spoke to him at a later point, I learned that Paul had indeed left and was seeking a more understanding and tolerant medical facility.

American bioethicists debate whether physician-assisted suicide straddles the theoretical bright line, but most agree that active euthanasia—the administration of a lethal substance, such as potassium chloride or carbon monoxide—crosses it. In all the issues surrounding end-of-life care, there is perhaps no name more controversial than that of Dr. Jack Kevorkian. Any discussion about assisted suicide or active euthanasia invariably polarizes people, but such conversations are further inflamed by mentioning Kevorkian's publicity-seeking behavior and fanatic crusade. When this subject arises, I have witnessed some of the most thoughtful and balanced people in the field of palliative medicine instantly regress, bicker, and rudely interrupt each other.

It is not commonly known that Kevorkian routinely videotaped his patient interviews. Selections from the videotapes have been broadcast on television, and in those recordings he comes across as simultaneously fascinating and repulsive. My own reactions are difficult to reconcile—I admire his assuredness but would never wish to make his acquaintance. Kevorkian is a folk hero to numerous people, and a Google search uncovers hundreds of related websites, including a trove of Kevorkian-inspired cartoons and a musical group called Dr. Kevorkian and the Suicide Machine. No less august a star than Al Pacino has been cast in the lead role of Barry Levinson's film, *You Don't Know Jack*, about Dr. Kevorkian. Kevorkian is also the protagonist in a tiny book published by Kim's favorite author, Kurt Vonnegut, entitled *God Bless You, Dr. Kevorkian*.

For several years leading up to June 2007, "Dr. Death," a retired pathologist, resided in a six-by-ten-foot jail cell at a Michigan prison.

Forbidden to talk to the press, he was serving an eleven-to-twenty-year sentence for the second-degree murder of Thomas Youck. Kevorkian injected Youck, a fifty-two-year-old Michigan accountant who suffered from amyotrophic lateral sclerosis (Lou Gehrig's disease), with a fatal combination of three medicines.

Kevorkian had arranged for a videotape of his final meeting with Youck to be shown on the CBS program *60 Minutes*. At the sentencing following his trial, Judge Jessica Cooper said, "You had the audacity to go on national television, show the world what you did and dare the legal system to stop you. Well, sir, consider yourself stopped."

Youck's death marked the fifth time that Kevorkian had been indicted, and it was the only time he was ever accused of first-degree murder. He had previously been acquitted by three juries for assisted suicide and a mistrial was declared in the fourth case. Incidentally, advocates of physician-assisted suicide prefer to substitute the phrase *assisted dying* for the emotionally evocative and pejorative word *suicide*. In contrast to his more usual practice of providing an apparatus with which individuals could administer lethal substances to themselves, Kevorkian recorded himself injecting Youck—a clear example of stepping over the bright line and performing active euthanasia. Kevorkian further compounded matters by dismissing his lawyer and insisting on acting as his own attorney.

Invariably the families of Kevorkian's so-called victims have been his biggest supporters, and their testimony has been crucial in his acquittals. However, lacking sufficient legal knowledge to respond to the prosecution's objections, Kevorkian was unable to have Youck's widow testify in his trial. She later told reporters that if given the opportunity to address the jury, she would have praised Kevorkian and explained that he had openly carried out the wishes of her terminally ill husband. Following the guilty verdict, Youck's widow and her brother later joined Kevorkian and other supporters for a subdued dinner at a friend's home.

In contrast to the families, the medical establishment has largely opposed and reviled Kevorkian's participation in what may be as many as 130 deaths. A spokesperson for the American Medical Association has stated, "By invading the physician-patient relationship to cloak his activities, Jack Kevorkian perverts the idea of the caring and committed physician, and weakens the public's trust in the medical profession."

The public does not entirely agree. A *60 Minutes* poll taken directly following the broadcast reported that 49 percent of Michigan respondents felt that Dr. Kevorkian had acted appropriately with Youck, 41 percent felt that he had acted inappropriately, and 10 percent were undecided. A CNN poll conducted twenty-four hours after the sentencing found that 67 percent of the nearly 72,000 people in an online survey believed the verdict was far too harsh. Paroled on condition that he not participate in any more deaths, Kevorkian is now limited to the speaker circuit and has been occasionally appearing on college campuses and television shows. The movie about his life will again thrust him into the spotlight.

While active euthanasia—such as what Kevorkian did to Youck—involves deliberate actions such as a lethal injection to end a patient's life, the term *passive euthanasia* is used to describe ending life through omission more than commission. I did not initially appreciate that, especially from an international perspective, the practice of treatment withdrawal—and to a lesser extent withholding treatment—is considered by many authorities to be subsumed under the rubric of passive euthanasia. The overwhelmingly positive American view of withdrawal and withholding—with the emphasis on caregivers' motivation to relieve suffering and not prolong the process of dying—is blotted out in many other countries around the world by use of the word *euthanasia*. The E-word—whether or not modified by *active* or *passive*—carries with it a powerfully negative emotional valence that leads many people to an exaggerated reaction or overreaction.

The ethical and legal stance of the United States on this matter holds sway in North America, much of Western Europe, parts of Eastern Europe, Australia, and New Zealand. In those regions it is agreed that when a patient is deemed to be "nonsalvageable" and the individual or his or her surrogates prefer to end life-prolonging treatments, then doctors will issue a do-not-resuscitate order, remove a ventilator, stop vasopressor medications that control blood pressure, and discontinue the various types of artificial nutrition and hydration.

There are large swaths of the world, however, that have different views on such matters. This appears to be the case in Asia, although it is difficult to generalize because its various countries appear to be developing highly individualized approaches to end-of-life care. Japan, for example, has recently experienced a lowering of physicians' status in society, and medicine is in the process of becoming more akin to a service industry. Accordingly, the Japanese patient and family are considered to be in a relatively elevated position where they can demand even futile treatment. Also, in Japan, China, and other Asian countries there is a much greater cultural reliance on body language instead of explicit verbal communication. Words are considered to be extremely potent; it would be unusual for a doctor to directly communicate a poor prognosis to a patient and to engage in a meaningful discussion of terminal care preferences. Since dying is not directly acknowledged and discussions between Asian patients, families, and health care professionals are rare, this may effectively stymie consensual end-of-life treatment decision making—the ideal in Western palliative medicine.

A recent report from Turkey—a country that bridges Asia and Europe—underscores the complexity of communication. The study found that while 90 percent of a sample of Turkish nurses would like to inform patients of their medical situations, one-third of patients did not want to be told the correct diagnosis. When Turkish patients expressed a desire to learn about their condition, they preferred that this communication take place only when family members were present. They also explicitly wanted their doctor to be broadly smiling while conveying any dire findings and recommendations.

In Israel, as in many countries around the world, there has been an obvious and long-standing clash between secular liberals and religious nationalists that impacts medical practice. Israeli medicine—which is both heavily influenced by Orthodox Judaism and follows a strong Hippocratic pro-life tradition—has not followed America and the United Kingdom's lead in dramatically lowering the threshold for treatment termination. An Israeli intensive care unit study has demonstrated that their intensivists only withhold and generally do not withdraw life-sustaining treatments. In other words, clinicians can refrain from starting life support, but once treatment is begun it is not stopped until the patient has recuperated or died. Although there are exceptions, Israeli medical practice dictates that even in the context of clearly defined terminal disease the prolongation of life is required; however, an act of omission, such as withholding treatment, is interpreted by the rabbis to be passive noninterference with the process of dying and is permissible, as are some other steps that can be taken to curtail suffering.

In December 2005, following nearly five years of intense deliberations, Israel enacted a new law, the Patient Nearing Death Act (also translated as the Dying Patient Act). The ruling is quite unlike anything we have in the United States, and it represents a consensus—or more accurately a compromise—hammered out by the secularists and the religious conservatives. The framework is predicated on balancing three fundamental values: the importance of quality of life, the autonomy of human will, and the sanctity of life. (This latter term does not exist in traditional rabbinic Judaism but was adopted for use in the act.)

The law came about because the Israeli medical establishment had an intense reaction to a case in which a patient with amyotrophic lateral sclerosis (the same disease as Kevorkian's patient Thomas Youck) was taken off a ventilator. A task force was formed consisting of five subcommittees and comprising fifty-nine physicians, ethicists, attorneys, and religious authorities (Jewish, Christian, Muslim, and Druse). This approach stands in contrast to the American process of relying more heavily on the judiciary to establish precedent.

Professor Avraham Steinberg organized the task force, and Professor Charles Sprung was chairman of the medical subcommittee that drafted the act; by coincidence, the latter was my medical school roommate. According to Charlie, the law is perhaps the first in the world to explicitly state that it is a citizen's right to have palliative care. This means that the patient and family should not have to suffer at the end of life and are entitled to have access to expert assistance for symptom management. A number of the legislation's provisions are equally foresighted, including a legal requirement that physicians inform terminally ill patients about their condition (if they want this information) and encourage patients to express preferences regarding medical care.

In addition, the Ministry of Health established the Central Data Base of Advance Medical Directives to update people's living wills and choices of health care proxies every five years. This information will be available to clinicians and hospitals to assist in terminal care. The law states that typical patients are assumed to want to live, and in the absence of evidence to the contrary they are to be provided with life-sustaining treatments. That presumption is reversed, however, when multiple organ failure is occurring *and* significant suffering is present *and* life expectancy is less than two weeks. Only when all three of these conditions are present does the individual then qualify for potentially stopping curative efforts and instead relying solely on palliative treatments. Israel differentiates between terminal illness, in which someone is expected to die within six months (the same criteria as in the United States), and imminent death, which means within a couple of weeks.

Charlie is an observant Jew. The bright red hair that I remember has now turned gray, but he still wears a crocheted yarmulke on his head. During his lengthy career as the medical director of the general intensive care unit at Hadassah Hospital, he has never discontinued life support from a terminally ill patient. However, the Patient Nearing Death Act offers novel ways to potentially work within the withdrawal prohibition. To begin with, the law carefully differentiates between "cyclical" and "continuous" treatments. Since hemodialysis is only conducted three days a week for several hours at a time, it is considered a "cyclical" therapy and it is permissible to discontinue such a treatment. By contrast, ventilators are a "continuous" therapy and discontinuation is prohibited. However, even here the law offers a loophole: the Ministry of Health has designed and is currently testing special ventilators containing timing devices that periodically issue audible warning signals and automatically shut down after a designated amount of time. When the alarm is sounding, physicians are then in a position to "withhold" rather than "withdraw" treatment from patients, such as those with ALS who are not clinically responding and want the ventilation to end. Incidentally, this is not an entirely new idea in Israeli medicine, as it was suggested by one of their celebrated rabbinical scholars thirty years ago for use at a religious hospital. Another loophole is currently available for patients maintained on vasodilators: physicians will not stop the medication but can instead avoid increasing the dosage as the blood pressure begins to fall.

Another facet of the Israeli law has to do with "basic care," which includes artificial feeding, insulin injections, and so forth. Generally it must be administered to incapacitated patients even if they have previously stated in an advance directive or living will that they do not desire such measures (there are again some exceptions). Between this last provision and the requirement that a rapidly terminal illness be present, Israel's position is diametrically opposite to the rulings of the decisive U.S. Supreme Court cases that currently shape American end-of-life medical care. Neither Karen Ann Quinlan, Nancy Cruzan, nor Terri Schiavo, all of whom were in persistent vegetative states and were the subject of litigation, would have had nutrition and hydration and other life support stopped under the Israeli law. I suspect that Olga Vasquez, Bobby Schindler, and the members of the Euthanasia Prevention Coalition will each find a great deal to approve of with the Patient Nearing Death Act. Interestingly, the Israeli legislature also firmly closes the door for palliative care murder accusations, as any possible consequences of violating the act's provisions will entail civil and not criminal prosecution.

Europe remains a continent that is deeply divided over how to manage the terminally ill. In 2002, the Netherlands was the first country in the world to legalize active euthanasia—although it had quietly tolerated the procedure since the early 1970s. The current rules are restricted to patients with incurable conditions who face intolerable suffering and who explicitly request euthanasia, although critics of this system point out instances in which the rules are not followed. Dr. Rob Jonquière, chief executive officer of the largest Dutch right-to-die organization, which has more than 100,000 members, told me in his Amsterdam office, "The request [for euthanasia] is so crucial in our law, in our culture, in our procedures, that if there is no request we never talk about euthanasia."

Jonquière attributes the birth of his organization and acceptance of all forms of euthanasia in the country to a widespread reaction to the case of Dr. Geertruida Postma. In 1971, Postma injected her own mother with morphine and curare. On numerous prior occasions, the two women had had end-of-life discussions and the mother repeatedly expressed her wish for assistance in terminating life if she ever became incapacitated. The mother then suffered a brain hemorrhage and was confined to a nursing facility; she could hardly speak, hear, or sit up. Postma and other Dutch physicians were accustomed on rare occasions to privately accede to euthanasia requests, and Postma therefore did not anticipate legal difficulties when she informed the nursing home's medical director of her intention to comply with her mother's wish. However, she was subsequently charged with murder. According to Dr. Jonquière, the presiding judge is reported to have said in confidence, "If I am in that situation, I only hope I will have a doctor like her!"

In 1973, the Leeuwarden criminal court found Postma guilty, but it rendered a token punishment consisting of a one-week suspended sentence and one year's probation. The Royal Dutch Medical Association then issued a statement arguing that the administration of pain-relieving drugs and the withholding or withdrawal of life support in futile situations were justified even if death resulted. According to Jonquière, the unrelenting clamor of the Dutch public over the next thirty years that Postma "was not a criminal and certainly not a murderer" directly resulted in the Netherlands taking the bold step of legalizing active euthanasia.

Later in 2002, Belgium became the second country from the European Union to legalize this practice. Under Belgian law, a person must be "capable and conscious" and able to request the procedure in a "willing, thoughtful, and repeated" manner. In order to avoid prosecution, physicians must be sure the patient "is in a terminal medical situation [and enduring] constant and unbearable physical or psychological pain" from an accident or incurable illness.

Perhaps the most highly publicized Belgian case took place in 2008 and involved the celebrated author Hugo Claus. Claus made his request after developing Alzheimer's disease, and the Flemish minister of culture later stated, "I knew him well enough to know that he wanted to depart with pride and dignity." The death was criticized by the Roman Catholic Church and the Belgian Alzheimer

League—the latter respected the decision but felt the media gave insufficient attention to other options available to people who develop dementia.

In February 2009, amidst heated public debate, Luxembourg likewise legalized active euthanasia and assisted suicide. Euthanasia is to be permitted for the terminally ill and those with incurable diseases or conditions only when they have repeatedly asked to die and have obtained the consent of two doctors and a panel of experts. The Catholic Church and most formal medical organizations argued against the legislation, but Luxembourg's citizens are renowned for their agnostic attitude. Jean Huss, a member of the Green Party in Parliament and cosponsor of the bill, said that "the Christian Social People's Party [of Prime Minister Jean-Claude Juncker] and the Catholic church were against the euthanasia law, calling it murder; but we said no, it's just another way to go." Luxembourg is a small country wedged between Belgium, France, and Germany. The majority of its citizens and legislators felt so strongly in favor of the measure that when Grand Duke Henri refused to sign the bill, the country's constitution was changed to make Henri's signature unnecessary. Calls have now arisen to make further sweeping changes to the constitution and remove more of the monarch's executive powers.

In the rest of Europe active euthanasia remains illegal, and it is passive euthanasia that engenders controversy. In France the unsuccessful prosecution of a mother and doctor for terminating the life of a quadriplegic man has recently led to an end-of-life law that advises physicians to avoid taking extreme measures to keep dying or brain-dead patients alive. In the city of Dijon, a court case that sanctioned withdrawal and withholding of treatment has been surrounded by heated public debate.

In February 2009, Italian prime minister Silvio Berlusconi similarly jumped into an emotional right-to-die debate around Eluana Englaro, who has been called Italy's Terri Schiavo. The thirty-eight-year-old Englaro had been in a persistent vegetative state since a car crash in 1992, and after a ruling by Italy's top court her father and doctors finally disconnected her feeding tube. This took place at a hospice in the northern city of Udine, the only one in Italy that had agreed to implement the decision to stop feeding and hydration. Berlusconi later announced, "Eluana did not die a natural death, she was killed."

The death occurred while the Senate was in the midst of debating a law that would have forced the clinic to replace the tube. Following the announcement, a moment of silence was observed. It was then broken by raucous lawmakers who began screaming at each other, and the word *murderers* was repeatedly shouted. Javier Cardinal Lozano Barragán, the Vatican's equivalent of a health minister, said in an interview, "To withdraw food and water from her means only one thing, and that is deliberately killing her. . . . May the Lord forgive those who brought her to this point." For three days in succession, Pope Benedict XVI indirectly referred to the case, at one point telling the Brazilian ambassador to the Vatican that "the sanctity of life must be safeguarded from conception to its natural end." A new law about end-of-life care in Italy will almost certainly result. Pundits suggest that it will likely condone withholding and withdrawal of life support treatments, but it is unclear whether this will specifically encompass the removal of artificial nutrition and hydration.

Spain, another Mediterranean and predominantly Catholic country, is racked by controversy over its end-of-life practices. There are no specific laws that regulate withdrawal or withholding, and most decisions take place in the context of confidential doctor-patient relationships. In 2007, a fully competent patient who was permanently dependent on a ventilator requested that it be discontinued. At the time she was residing in a hospital owned by the church, and the ecclesiastic hierarchy decided to transfer her to a public hospital, where the ventilatory support was then withdrawn. This was not labeled as euthanasia—passive or otherwise—and there were no legal ramifications, although the incident was widely argued in the press.

The Spanish physicians whom I have spoken to stress that forgoing and terminating life support treatments, along with palliative sedation, active euthanasia, and assisted dying, are presently being publicized for political purposes by different parties. In early 2008, a high-profile case involving fifteen physicians alleged to have caused premature deaths through oversedation of four hundred terminally ill patients at a single hospital was dismissed after a lengthy and rancorous investigation. All of the accused physicians were fully absolved.

The Swedish Society of Medicine reassessed its rules after a thirty-five-year-old ventilator-dependent citizen was unable to persuade doctors to withdraw life support and was then compelled to travel to Switzerland, where assisted suicide is legal. Sweden now urges physicians to respect the terminal treatment preferences of patients who retain the capacity to assess their clinical situation. Active euthanasia remains illegal and is not favored by most Swedish doctors.

The Swedish man who sought to die in Switzerland is but one of hundreds of Europeans who have traveled to Zurich seeking the aid of Dignitas. This volunteer organization was instituted in 1998 to help people with terminal illnesses, and it provides instructions and barbiturates for suicide. It is not wholeheartedly embraced by the Swiss, and the headquarters of the organization has had to move around to different addresses because of opposition from local residents—reminiscent of the hostile reception of methadone maintenance clinics in the United States.

Many Germans have sought the assistance of Dignitas, and yet for obvious historical reasons euthanasia remains a highly taboo subject in that country. Articles suggest that German medical staff are especially uncertain about the ethics and legality of all types of end-of-life procedures. Assisted suicide, for example, is no longer technically illegal, but it cannot involve physicians. This year, one of my favorite titles for a medical article was "A 'Little Bit Illegal'? Withholding and Withdrawing of Mechanical Ventilation in the Eyes of German Intensive Care Physicians." This qualitative study concluded that in Germany the fear of making unjustified or illegal decisions is paralyzing medical care and motivating doctors to continue even futile treatment.

For more than twenty-five years the United Kingdom has struggled to chart out an acceptable end-of-life policy. In 1981, Dr. Leonard Arthur, a highly respected English pediatrician, was charged with murder following the death of a baby with Down syndrome. During his trial the charge was reduced to attempted murder, and Arthur was ultimately acquitted. For several years following this case, British physicians worried that withholding of life support would leave them liable to allegations of murder; treatment withdrawal was considered by many doctors to be a safer course of action.

This remained the status until Anthony Bland, an avid supporter of the Liverpool soccer team, traveled to the Hillsborough soccer ground to watch a semifinal match and was caught in a melee involving his fellow spectators. Ninety-five people died in the crowd, and he sustained severe injuries that resulted in a persistent vegetative state—the same form of brain injury as Terri Schiavo. Transferred to the care of Dr. J. G. Howe, a consultant geriatrician and neurologist, the twenty-two-year-old man was kept alive by artificial nutrition and hydration and skilled nursing care.

In 1989, Howe contacted the coroner and informed him of a "plan to withdraw all treatment including artificial nutrition and hydration." Unlike the dissension surrounding Terri Schiavo, the decision concerning Anthony Bland was reached in full accordance with his family's wishes.

Howe has since written, "It is difficult now [in 2006] to convey my shock on receiving his intimidating reply. Having stated that he [the coroner] had no jurisdiction over any living person, he advised that I would risk a murder charge should I withdraw treatment. He made it clear that he . . . could not countenance, condone, approve or give consent to any action or inaction which could be, or could be construed as being, designed or intended to shorten or terminate the life of this young man. This particularly applies to the withholding of the necessities of life, such as food and drink."

Dr. Howe was visited the next day by the police, who reiterated that if he withdrew treatment and his patient died, murder charges would be brought. The Airedale National Health Service Trust applied on behalf of the physician and the family to the courts, and then dealt with an appeal. The courts finally determined that withdrawal of treatment was a reasonable course of action and murder charges would be highly inappropriate.

In the United Kingdom, like the United States, withdrawal and withholding of life support treatments are now widespread and considered to be both legal and ethical. This is not to say that accusations and bitter investigations aren't occurring, particularly as related to discontinuation of life support therapies. A British nephrologist colleague is currently undergoing hospital disciplinary procedures and has been suspended from clinical duties following two separate cases in which she wanted to stop dialysis but was opposed by the patients' adult children. In both instances, the nephrologist spoke to the relatives and explained that the patient was dying, continuing dialysis was futile, and there was a substantial risk of death ensuing during a dialysis session. She attempted to explain that the appropriate treatment would be to keep the patients comfortable and enable the family to be present for a dignified death.

The first patient was Somali, and the relatives threatened to go to the press and say the hospital personnel were racist. When word of this got back to the medical director, she promptly ordered the dialysis restarted. The patient died two weeks later while in the midst of a dialysis treatment. Dialysis was also resumed for the second patient, and he died one hour later while hooked up to the dialysis machine.

In England, the expression "conservative treatment" is commonly used to describe an approach to renal failure that entails refraining from initiating dialysis for self-selected patients. There are clinical programs in the United Kingdom (but not the United States) that currently provide these patients with comprehensive palliative care and support them and their families in the resolve to not begin dialysis.

In October 2009, new rules were issued in England and Wales that clarify when prosecutions for assisted suicide are to take place. The rules followed a case in the House of Lords in which a woman suffering from multiple sclerosis who wanted to seek the aid of Switzerland's Dignitas organization requested clarification as to whether her spouse would be subsequently prosecuted. The guidelines take into account the dying wishes of the deceased, their age, presence of mental illness, and capacity to make the decision, as well as whether family or friends stand to benefit from the estate or are primarily motivated by compassion. Assisted suicide remains a crime in the United Kingdom, but the decision to launch criminal proceedings is not automatic and will now be influenced by thirteen factors against prosecution and sixteen for prosecution.

Mexico is also struggling over the issue of terminal care. For example, in April 2008 the Mexican Senate voted 70–0 in favor of legalizing passive euthanasia. The law would explicitly permit physicians to withdraw life support treatment and provide medications for pain control when patients have a six-month prognosis and are receiving palliative care. The bill still needs to be ratified by the Chamber of

Deputies. It was drafted in reaction to Mexico's previous policy of prosecuting physicians who provide end-of-life care and the country's unfortunate tradition of demanding lengthy jail time for anyone who aids patients in ending their lives.

In August 2009, Western Australia's highest judge, Wayne Martin, said a nursing home would not be criminally responsible if it stopped feeding and hydrating a quadriplegic man and thereby facilitated his death. Christian Rossiter had become paralyzed following a traumatic injury, and the judge ruled that although he was not terminally ill, Rossiter retained the capacity to make his request. The ruling sets a legal precedent in Australia, where assisting someone to take his or her own life can be punishable by life in prison.

Bulgaria is representative of the problems of palliative care in the many countries that cannot afford expensive medical technologies. When I visited Bulgaria for the first time in 2001, it was immediately obvious that my planned presentation on the issue of dialysis discontinuation would have little relevance. I spoke at the medical school in Sofia, where plaster from disintegrating walls and ceilings littered the floor; the stairwells were dark because the lightbulbs had been stolen. Dialysis was far too expensive a therapy to be widely available. Of greater import to my audience were the topics of patient autonomy, medical paternalism, and the disclosure of honest diagnoses to patients. A couple of the medical students in the audience brought to my attention that there were at least two patients presently on the pulmonary ward with metastatic lung cancer who had been reassured by their physicians that they merely had bronchial infections.

Not surprisingly, many of Bulgaria's most ambitious but impoverished medical personnel have looked for professional opportunities in other countries. In 1999, nineteen foreign medical workers working at a Libyan hospital were accused of infecting hundreds of children with the AIDS virus. This is now recognized to have been the most blatantly xenophobic example of murder accusations involving medical personnel. Thirteen of the individuals were released, but the remaining five Bulgarian nurses and a Palestinian physician were tortured, tried, found guilty of murder, and sentenced to death. In July 2008, after considerable negotiations on the part of the European Union and the United States—and a $1 million payment to each of the families of the infected children—Libya's highest judicial council commuted the sentences of the six medical workers to life in prison. This was followed a week later by the successful intervention of Cécilia Sarkozy, then France's First Lady, who persuaded Colonel Muammar el-Qaddafi to free all of the prisoners.

Looking at this from a global perspective, it is evident that the United States is hardly alone in its difficulty adjusting to the advances in medical technology and the need to design humane policies related to dying. As described above, there have been numerous instances throughout the world involving the criminal prosecution of health care professionals for passively or actively helping patients to die. Many of these cases appear to precipitate a societal debate that not infrequently leads to the national legislature liberalizing and legalizing medical practices. In three European countries, not only do laws now safeguard withdrawal or withholding of life support treatment and the vigorous provision of analgesics, but the Rubicon of the bright line has been definitively crossed and active euthanasia legalized.

Nevertheless, the ambiguities continue, and caught in the crossfire between politicians, religious leaders, bioethicists, law enforcement officials, and lawyers are the doctors and nurses—the people like Amy and Kim—who have to participate in these crucial decisions every day. Each country has its own versions of cases like Terri Schiavo's or zealots like Dr. Kevorkian, and though these high-profile situations spur conversations around the world, the subsequent urgency of the issues gets lost in the tumult of the debate. Meanwhile for Amy, Kim, and their counterparts, there's little time or patience for ambiguity; as long as a patient's suffering is on the line, they are going to be doing whatever they can to alleviate it—and hope that they won't become targets of those who disagree.

16

Sudden Awakening

As Amy and I sat in the restaurant, other patrons were relaxing to the music or engaged in light banter about friends or their children's latest scrapes or accomplishments. Amy and I were instead caught up in the horror of her experience at the Massachusetts State Police barracks.

Amy dropped her fork. She looked around anxiously, concerned that another diner might have witnessed her being clumsy. As best as I could tell, the only one who noticed was our attentive waitress, who popped over with fresh silverware.

"They question me for almost four hours," she exclaimed with some exasperation. "I still have no idea what we are doing. The police never say, 'You have been accused of murder.' Instead, they are vague about everything, and they keep asking me questions about the TA, Olga. Questions like 'Does Olga have a grudge against you?' or 'Is there some reason that she would be doing this?' Then it dawns on me what is going on, and I literally shout, 'Are you telling me that Olga said we killed this woman?'

"In retrospect, I think the detectives must have been thinking, 'No one who actually killed somebody could be as stupid as this girl!' I was just so oblivious. I could not believe anyone would reach that conclusion. I am convinced that being ignorant totally saved me, because they probably thought I never could have made up this whole story. I mean, I spelled things out for them. I could not have been more cooperative."

Amy's napkin, which she had been twisting into knots, slipped off her lap and joined the errant fork on the floor.

She continued, "I quietly tell the police that I had exchanged Christmas presents with Olga on Sunday morning. I was her Secret Santa. After all, it was that time of year. I think we are friends."

Amy looked up from the table and turned toward me. "I knew she was religious and belonged to a church called the Assembly of God. I didn't know what Assembly of God people believe or do not believe in. I didn't know if it was a religious thing and she had an issue with it. Plenty of our patients have died, and Olga had helped us to make them die comfortably. It wasn't that this was her first death and she was freaking out. I have barely spoken to Olga since that day, but I guess in her head, Kim and I are killers. Apparently she and her lawyer convinced the district attorney, Bill Bennett, we murdered this patient, and that is where everything started."

Amy took a sip of water and said, "After hours of going through it a million times and asking me the same questions fifty thousand different ways to see if I am changing my story, the state police detectives let me go. It has been a long evening and I have been entirely honest. I am exhausted by the process of describing my role and defending Kim's integrity. It is not until I am in my car driving home when I fully appreciated that I am on the hook with Kim. It is midnight now and I call my husband, but I am totally paranoid and will not tell him anything. I am convinced that my cell phone is being tapped. I want to warn Kim, because they will be driving to her house. But I am afraid to talk. How can I tell Kim what is going on?"

When Amy got back to her house, she broke down in her living room. After a while the phone rang. It was Kim. She had just gotten a call from the state troopers telling her that they were en route. Apparently they already knew that Kim was a single mom, and so they couldn't take her downtown. Still, Kim insisted that she wouldn't allow them into her house for questioning if they arrived in an unmarked car; she explained that she lived out in the middle of nowhere, and they could come only if they drove a clearly marked police car.

After minutes of frantic chattering back and forth on the phone, Kim said, "Amy, what is happening?"

"I don't want to tell you anything," Amy began, "except this is the absolute worst experience of my entire life, and unfortunately you have to go through it, too. . . . Just answer their questions and be perfectly honest."

Shortly afterward, a state police cruiser pulled up in front of Kim's house. The two detectives emerged, everyone went into the living room, and they proceeded to question her until early morning.

Kim's account of the evening's events and her encounter with the state police paralleled Amy's.

"I had worked a regular shift, gone home, and climbed into bed," she explained. "About eleven or eleven-thirty P.M. the phone rings and it is the police. They say the coroner has some questions about Rosie's death, because she was so young. They choose their words carefully, but they are basically lying. They want to come to my house to talk about it. And of course, I want to do whatever I can to help. I say, 'Okay, okay, sure, come on over.' Heck, I just worked twelve hours, and pretty soon my kids have to get up and go to school, and I have to go to work the next day, but yeah, let them come."

And that was when Kim called Amy. After they got off the phone, Kim's anxiety set in. All she could do was wait.

"I march into the garage and smoke thirty-seven cigarettes in about two seconds. I come back in the house and my ex-husband— he was over for a visit at the time—says, 'Relax. They are just going to talk to you about, you know, whatever. No big deal.' My heart is pounding, and I am thinking, 'Something is not right. Something is not right.'

"It is midnight, the police arrive, and they come in, sit down, and start questioning me. They ask, 'What is dialysis?' and 'What is this?' and 'What is that?' They are going on and on and on. I am completely exhausted. I have given them every medical lesson I can, because every time I say something, they go, 'What?' I explain renal failure, I explain bedsores. I explain chronic obstructive pulmonary disease [emphysema] and why you should not use supplementary oxygen. Blah, blah, blah. 'My God,' I think to myself, 'these are the stupidest men I have ever met.'

"So finally, at about three in the morning, they start asking questions about morphine, like 'What constitutes a safe dose?' and 'How much do you usually give?' and 'How much would it take to kill somebody?' I am half asleep and reclining on my couch, when all of a sudden it hits me and I sit bolt upright."

As she said this, Kim looked across the table at me, dramatically mimicking her sudden awakening and abrupt loss of innocence. Her eyes bugged out and her mouth dropped open. Kim's voice apparently cracked when she confronted the detectives, and she squeaked at them, "I did not overdose that woman. I only gave her what the orders called for me to give her."

"It just hit me—like bang!" she recounted, clapping her hands explosively. The people at the surrounding tables in the restaurant were startled and glanced at us. "The detectives think I did those things. Up until then, I thought I was just being a good little helper. What had I fallen into? Oh, my God!"

The detectives kept asking Kim questions until around five in the morning. When they were done, they pointed at the screen of the laptop and said, "We are going to print this and have you sign it."

Kim had been reduced to exhausted mumbling, so she told them, "Whatever you want me to do, sir," and "Okay, sir."

Rolling her eyes at me, she said, "Then their portable printer malfunctions, and, being a good little criminal, I say, 'Here is a clean disk. Save it to the disk. We can use my computer and printer.' At last the file is printed, I sign it, and the state troopers depart.

"I go upstairs and try to sleep. But I cannot fall asleep. I keep thinking, 'Oh, my God, they believe I did something wrong to this lady. Oh, my God! Oh, my God!'"

At around seven in the morning, Kim called Amy again.

"What do they think we did?" Kim asked.

"I don't know, but I guess they think we killed her" was the only response that Amy could muster.

"You have got to be kidding me! What do we do now?"

"I don't know."

"Should we call the hospital?" Kim asked.

"Well, who are we going to call at the hospital?"

"I don't know!"

Finally they decided to call Eileen Grunwald, the acting director of their unit. When Kim called, Eileen was not in yet, so Kim left a message, and shortly thereafter Eileen called back, asking Kim to come down to the hospital to talk about what had happened.

"Eileen, I did not kill anybody. I swear to God."

It was clear to me as I sat across from Kim in the restaurant that the emotions of that day years earlier were still incredibly raw. She slowly moved the food around her plate and looked up at me.

"I was beside myself. I was thinking, 'All these years, all this time, all I have ever wanted to do is help people, and it is blowing up in my face. I did something wrong but cannot for the life of me figure what it is that I have done.' I keep thinking, 'Maybe I grabbed the ten-milligram vial of morphine instead of a two-milligram vial?' I cannot remember doing anything incorrectly. I just cannot.

"At midmorning I arrive at our meeting, and Eileen explains that accusations have been made that I overdosed Rosie. Olga has told the police a bunch of things. She has accused me of giving medicine without orders, and she has told them that I tried to kill Rosie faster by removing her oxygen.

"The police conduct a search of the Pyxis machine [the computerized mechanism that dispenses medications] and they review the doctors' orders. The detectives identify a single discrepancy— and it is not really a discrepancy—that I wasted two milligrams of morphine without Amy watching. Back then, we all were very lax about watching each other waste. We would cosign the waste and would trust our colleagues. This subject has come up during my interrogation, and I told the police that Amy did not watch me squirt the excess. And earlier in the evening, Amy had apparently told the police the same thing. In the end, the accusation that I overdosed Rosie was based on a grand total of two milligrams of morphine."

"Let me just get clear on this," I said to Kim. "Was there a reason you used a four-milligram vial rather than a two-milligram one?"

"Because at the time I was hoping that the intern would call and change the orders," she replied. "But unfortunately that did not happen."

Kim continued to talk about the subsequent hospital meeting. "It turns out that all my orders are correct, and what I charted as having been administered completely correlates with the computer record. I have correctly documented everything, and the doctors support my account. When the physicians are interviewed about possible deviations from the orders, they all say, 'No, she has done exactly what she was supposed to do. She is also completely right on the money when she asked for more medicine.' They are totally supportive. The hospital peer review committee sits down with the police, and Dr. O'Shea [a nephrologist on the committee] laughs when they are questioning him about the morphine, because I am clearly not giving big doses and the discrepancy is a minuscule two milligrams.

"Another thing," Kim exclaimed, "Olga also apparently informs the police that I was telling the patient it is all right to die and I was helping her. Olga is correct! I tried to help Rosie. At the time, I was thinking, 'She is dying, and maybe she is scared. Perhaps she needs to know it is okay that she is dying. I do not want her to be afraid.' I do not know what it feels like to be dying. None of us really knows what it feels like, because once you are dead—that's it! I mean, you go to the dentist for the first time, you are afraid. You go to the doctor for the first visit, you are afraid. The first time you drive the car by yourself in the rain, you are afraid. So my saying to Rosie, 'Don't be afraid, I'm helping you,' does not seem to me to be a particularly bad thing."

I asked Kim if she still talks to her dying patients. "I am a lot quieter when I say these things to people," she replies. "I whisper. I get very close to them, so that I know they can hear me. I think I am helping them to die comfortably with dignity and without fear."

For Amy, the time spent in limbo during the investigation was a nightmare.

"If I can get through the remainder of my nursing career and never have to do that again, I will be very happy, because it is absolutely dreadful," she told me, recalling that difficult time.

During her forced leave, Amy met repeatedly with the hospital's nursing director, the risk management staff, and a team of medical center lawyers. All the time, though, she said she worried most about her colleague. "It was worse for Kim," she remarked, "because she was directly accused. I was merely the coconspirator of the murder. Kim was looked at as being the actual murderer.

"Kim is younger than me, and she is a much newer nurse," Amy explained. "I remember one day in particular when Kim had a million terrible thoughts going through her head and was simply hysterical. Her head was on my lap, and she was crying. She was convinced that this was the end of her whole career, her profession, everything. Kim is a single mom who put herself through nursing school, and she sees her life completely crumbling.

"The more we think about what has happened, and the more we talk, the uglier it all seems. We feel as if we are already locked in prison. They will not let us work, because we are alleged murderers. How good would it look for the medical center if they continue to employ two nurses who are accused of killing a patient? Furthermore, this all happens at the same time as the Kristen Gilbert case, and it was surely not helping our situation."

Amy cleared her throat, then began to speak rapidly. "From one day to the next, I do not know if I am going to work or going to be indicted. Every day is a mystery. The Baystate people try to be supportive and helpful, but they are also powerless. Kim and I talk for hours on the phone each day, crying and worrying about our futures. I sit and stare at my Christmas tree as its needles and branches turn brown. I have no interest in watering it. I weep softly while the leaves drop off and form a pile on the floor. I wonder if I am going to prison. We are told not to talk to our coworkers, which is a problem because the coworkers are my friends. Finally, I say, 'Screw it! I am talking to them. I do not care. I have not been charged with anything, and I am talking to them.'

"The poor family of Mrs. Doherty is also affected by this situation. I understand that they have a wake and a funeral service, but the district attorney goes to the church and tells them that they cannot bury their mother. I do not know if it is the DA himself or somebody representing him, but the family is told that they have to release the body to the medical examiner, because of the murder investigation.

It is a really nice family with great kids, so I can only imagine what they think. It is another week before they are able to actually bury Mrs. Doherty's body. In the end, the family sends a fruit basket to the renal unit with a note thanking us for our wonderful care. I believe they appreciated us and our circumstances.

"The nursing director calls every day to see how we are doing, and we are invited to the hospital for occasional meetings with lawyers and members of risk management. During one such appointment, Kim and I are sitting down with a bunch of administrative people, and the phone rings. Somebody answers it, turns to the group, and says, 'The district attorney is in the building.'

"It becomes a scene from a horror movie. The phone rings, is picked up, and someone else announces, '*He* is now on the third floor.' Two minutes later, it rings again, and they report, '*He* is on the first floor.' Oh, my God, it is just awful. I do not know what *he* is looking for or who *he* is coming to see. All I know is that *he* and his minions are hunting for evidence that could hurt Kim and me. Staff from all over the hospital are having district attorney sightings and calling in reports. It is totally creepy."

Kim described her experience of being out on administrative leave. "I was a blubbering idiot. Sobbing, taking back everything I had said to Rosie. 'I should never have told her I was going to help! Why did I tell her it was okay to die? I should never have gotten that order for morphine changed—I should have just left it alone. What was I thinking?' I was second-guessing myself as a person, and the way that I have always practiced nursing. I could not get beyond the fact that this was like a bad movie. I was going to jail for the next thirty-five years and that was how it is going to be. Period. I was not going to be able to prove anything. I was going up to the Big House!" She glanced at me and laughed nervously.

"I did not sleep, and all that I consumed for the first couple of days was water. I smoked a whole lot. I thought I was having a heart attack several times a day. My body felt heavy and I couldn't catch my breath. It was unbelievable. A couple of times I wished I was Amy—then she would be the 'killer' and not me.

"On the positive side, I understand that when the police questioned the family about the death of their mother, they were entirely comfortable with the way things happened and with the nursing care. They like me, they were glad I was there, and they said I helped their mom."

All three nursing staff members were encouraged by administration to see therapists. Olga began counseling, and Kim saw a colleague of mine. Both of them found it to be helpful. Amy went for a single session. She had previously been in treatment around the time of her divorce.

"In case you are wondering," she told me with her usual grin, "I hate psychotherapy. I'm not a good therapy person. To be completely honest, I think it's the biggest waste of time on the face of the earth. I am also not a support group kind of girl. I will go to lunch with my friends. That is all the support I want. I am not going to sit with a stranger or a bunch of strangers and yabber about how somebody's actions ruined my life."

Olga described to me how she had flashbacks to the events around Rosie's final hospitalization. She explained the importance of rules in her life. "You must have structure," she told me. "And this has to come from your upbringing. We went through a lot as a family, but it was always based on honesty and truth. We didn't go to Catholic schools. We went to your normal public school in Harlem on 110th Street and Lenox. We were latchkey kids and had to hurry up and get home. There was no one at home waiting for us. We four kids had an honesty policy and we had to take care of each other. My father was in the merchant marine and we hardly got to see him. My mother worked two jobs at the factory and as a school aide. When we did get to see my parents it was pounded in that you had to walk the straight line. We were really respectful of our mother. She taught us that you have to be respectful—to others and to yourself."

As a child, Olga regularly attended church and sang in the choir. She had a typical teenage rebellion but emerged from it with enduring admiration for her mother and a conviction that "I will struggle to do good . . . dedicate myself to taking care of patients . . . and taking care of my family. It was my mom who made me into a strong, positive person."

In preparation for this book, Olga and I spoke several times on the phone, but she did not show up for a scheduled interview at a restaurant. I waited for a couple of hours, and she subsequently explained that there had been a miscommunication.

Eileen Grunwald is the acting director of the renal unit, and she is the person Kim and Amy decided to call when they found themselves accused of murder. She and Sharon Smith, the vice president for nursing at Baystate, are the administrators who worked behind the scenes throughout the investigation. The administrators did not tell Kim and Amy that they had been reported—probably by the district attorney's office—to the Massachusetts State Board of Nursing. The board sent the Baystate Nursing Service a letter formally requesting that Kim's license be suspended. Eileen and Sharon Smith (who has since died of cancer) resolutely refused this demand.

In their letter to the board, they wrote:

As we discussed on the phone, there was both an internal and an external investigation, including two extensive peer review audits that were conducted regarding the claim of a patient injury. We are pleased to report that the investigation(s) found no merit to the allegation. The medical staff involved in the care of the patient, as well as the hospital team, have stood solidly behind Nurse Kim Hoy throughout the period of inquiry.

The manager of Nurse Hoy's department has performed random narcotic audits for the last several years. Additionally, in response to the complaint, the manager has discussed proper narcotic wasting through individual conversations with all licensed nurses on the unit, as well as at a staff meeting. There have not been any issues regarding diversion, and the audits have not identified any narcotic discrepancies.

Nurse Hoy has been employed in good standing at Baystate Medical Center since June 1, 1992. She is viewed as a clinical leader, and is considered a strong advocate for her patients by her peers, manager, and physician colleagues. Her performance has either met or exceeded our expectations.

For three months Amy and Kim were out of work on paid leave. While the hospital continued to provide weekly paychecks, the nurses hardly felt grateful. They had each found themselves cast as Gregor Samsa in Kafka's "Metamorphosis," and it was an unendurable role.

17

War on Drugs

Prescribing and administering opiate analgesics such as morphine is complicated. Because America is vigorously waging a "war on drugs" and because these medications have abuse potential and a high street value, it is not surprising that their use would bring law enforcement and the field of medicine into conflict. When nurses and physicians are accused of murdering terminally ill patients, they are almost always alleged to have improperly prescribed opiates, engaged in poor record keeping, and of having themselves been drug addicts or drug traffickers.

The Baystate nurses were accused of incorrectly wasting excess morphine. This was also the original stimulus for the investigation of the aforementioned psychiatrist, Dr. Robert Weitzel. In fact, until his acquittal, the U.S. Drug Enforcement Administration (DEA) appears to have considered Weitzel to be the veritable poster boy of evil doctors. In 2002, then–DEA director Asa Hutchinson addressed the annual meeting of the American Pain Society; the text of his speech was posted on the DEA's website. The thrust of Hutchinson's argument is that the DEA trusts the judgment of physicians in prescribing narcotics and other controlled substances, and it does not intend to second-guess doctors when it comes to the treatment of pain. He makes the point that the vast majority of doctors will never encounter the DEA during their careers. In 2001, more than 900,000 physicians were registered with the DEA to handle controlled substances, but during that year the agency initiated only 861 investigations of physicians and took action against 697 violators. Most of these cases resulted in the surrender of DEA registrations *for cause*—a term signifying that the doctors were no longer entitled to a DEA number because they had been convicted of a drug-related felony or they had retired and were not licensed to practice medicine.

"So let me give you an example of the kind of doctors we do investigate," Hutchinson declared. "Dr. Robert Weitzel was a physician in Utah who was brought to our attention through an anonymous complaint. It turned out Dr. Weitzel was providing patients with prescriptions for the Schedule II opioids morphine and Demerol and requiring they return the drugs to him so he could administer partial quantities and keep the rest for his own use.

"He even picked up drugs himself at pharmacies that he issued in patients' names without their knowledge. Many of these patients never received the medications, and some had never been treated by the doctor. Dr. Weitzel surrendered his DEA registration and pled guilty to obtaining drugs by fraud.

"And so that illustrates the kind of doctors the DEA is targeting. Those who unlawfully deliver controlled substances. And let me be clear that the DEA will take strong enforcement efforts against people like Dr. Weitzel who are diverting controlled substances and who are causing such great harm."

When I asked Weitzel about these specific accusations, he calmly refuted each and every claim, maintaining that all of his patients were legitimately and professionally treated. As the attending psychiatrist in a headache clinic, he saw a number of individuals struggling with substance abuse, schizoaffective disorders, and dementias who he believes were vulnerable to browbeating by investigating agents. Robert continued to explicitly deny that he himself had ever had a substance abuse problem. He reiterated that his only legitimate misdeed was to practice in a clinic that did not fulfill the federal requirement of maintaining a countersigned record of wastage. I have no proof either way, but I believe him.

During a subsequent conversation with Dr. Lloyd Stanley Naramore, I was pleased to learn that following the conclusion of his legal troubles he was welcomed into a doctors' group that had organized a clinic for the treatment of pain in Cincinnati, Ohio. Which is why it was so discouraging then to hear that Stan had not only developed health problems requiring open-heart surgery but also had come under investigation by the Ohio State Board of Pharmacy. In June 2007, an affidavit from a compliance officer was filed as the basis for a seizure of his office computers, financial records, and personal correspondence. Stan was accused of prescribing narcotics to known drug addicts, charging cash for medical care, and creating a clinic that drew many of its patients from the adjacent state of Kentucky.

In response to a newspaper article and bloggers' comments, he posted an e-mail online that included some of the following comments:

> I am licensed by the Federal Government and hold a special Certificate from the Drug Enforcement Administration to treat drug addicts in a very private, confidential, and compassionate outpatient setting.

What kinds of patients see a physician who treats drug addicts? Drug addicts.

Do we have a lot of drug addicts in our practice that treats drug addicts? Of course we do. That's a tautology.

The Ohio Board of Pharmacy requested the medical records of some of our patients.

We provided those records. We will cooperate fully with any further requests.

I only hope that patients who were being treated for drug addiction and patients who were planning to be treated for their drug addiction will not now be forced to return to illegal and dangerous drug use, because of fear of the loss of their privacy.

The practice [consists of] myself and my assistant. If I were to accept insurances I would have to add at least two more people to my staff to bill the insurance companies, and charge patients much more, for no reason.

I do not have an attorney. I have no reason to think I need an attorney.

I believe in the immediate and adequate treatment of pain.

If a patient being treated for pain develops a dependence of any type, I believe he/she likewise should have immediate access to competent continued treatment for both their pain and their dependence. We have a long way to go to achieve that goal.

The under treatment of pain is a serious problem in America. Physicians are afraid of persecution and prosecution for adequately treating pain. No American should suffer pain. Nor should an American lose his right to privacy because he/she has the courage to seek treatment for addiction.

In June 2008, I contacted Stan again in order to get an update on his situation. He reported that his clinic had reopened, but he first had been required by the Board of Registration of Medicine to spend $5,000 for a four-day evaluation at a rehabilitation program in order to prove that he was not a substance abuser. Having accomplished this, he was placed on probation for the next two years. Stan concluded his e-mail by saying, "Carpe diem!"

I wish him well.

How big is the problem of prescription medication abuse? Worldwide, the abuse of prescribed drugs is about to exceed the use of illicit street narcotics. According to the 2006 annual report of the UN-affiliated International Narcotics Control Board, in parts of Europe, Africa, and South Asia prescription drug abuse has already outstripped traditional illegal drugs such as heroin, cocaine, and Ecstasy. In 2003, the number of Americans abusing prescription drugs had nearly doubled to 15.1 million, from 7.8 million in 1992. According to the report, although the number of U.S. high school and college students abusing illicit drugs declined in 2006 for a fourth consecutive year, "the high and increasing level of abuse of prescription drugs by both adolescents and adults is a serious cause for concern." College students' prescription drugs of choice include the painkillers oxycodone (OxyContin) and hydrocodone (Vicodin). The high potency of these synthetic narcotic drugs raises the likelihood of overdoses—and, as evidenced by the death of twenty-eight-year-old actor Heath Ledger, they are often unknowingly taken in fatal combinations.

A 2005 article in *Time* magazine entitled "Why Is the DEA Hounding This Doctor?" reported that in the previous six years, more than 5,600 physicians had been investigated and 450 prosecuted for illegal prescribing and drug diversion. While there were likely good-faith law enforcement objectives in play, the problem is that this activity leads physicians to develop fears of unwarranted scrutiny. Surveys from around the country have repeatedly found that doctors often purposely undertreat pain because they are frightened of triggering an investigation by regulators.

I am convinced that the overwhelming majority of doctors and nurse clinicians—particularly those employed in pain clinics and palliative medicine practices—are entirely well-meaning and blameless of any criminal intent. It is terrible to see innocent individuals targeted for providing analgesics when one estimate has suggested that there are as many as 75 million Americans suffering from chronic pain, and this situation results in more lost work days than from heart disease and cancer combined.

One of the problems faced by pain doctors is their support from organizations such as the Drug Policy Alliance (DPA). The DPA—as opposed to the DEA—wishes to create "a just society in which the use and regulation of drugs are grounded in science, compassion, health and human rights, in which people are no longer punished for what they put into their own bodies but only for crimes committed against others, and in which the fears, prejudices and punitive prohibitions of today are no more." In other words, the DPA embodies an ethos from the 1960s; it stands in opposition to the "war on drugs," the criminalization of marijuana, and limited access to syringes. Whether or not I agree with these sentiments, such an organization is not likely to be endearing to even a mildly conservative judge, lawmaker, prosecutor, or medical clinician.

Dr. William Hurwitz is a crusading pain doctor who thought he had finally come up with a system of safety checks that would permit him to practice without interference or accusations from the DEA or district attorney's office of the Commonwealth of Virginia. For many years, Billy, as he is known to many in the pain community, persisted—despite frequent clashes with authorities—in maintaining a large clinical practice where he espoused the use of high-dose narcotics. On April 14, 2005, he was sentenced to twenty-five years in prison for fifty separate counts of criminal drug distribution.

At a news conference, Karen Tandy, then head of the DEA, displayed a plastic bag containing hundreds of opiate capsules. John Tierney of the *New York Times* later wrote about the case in an article entitled "Juggling Figures, and Justice, in a Doctor's Trial." He quoted Tandy as saying, "Dr. Hurwitz prescribed 1,600 pills to one person to take in a single day. . . . [He was] no different from a cocaine or heroin dealer peddling poison on the street corner. . . . To the million doctors who legitimately prescribe narcotics to relieve patients' pain and suffering, you have nothing to fear from Dr. Hurwitz's prosecution." However, Tierney derided her assurances and concluded instead that physicians should be very afraid.

Billy's conviction was reversed by the U.S. Court of Appeals for the Fourth Circuit. The judge in the district court had erroneously instructed the jury that they could not consider whether he had acted in good faith in prescribing the opioids. During a subsequent retrial, Billy was cleared of most charges, including the one depicted in the show-and-tell publicity stunt where the prosecutor brandished the bag of opiate capsules.

Attorney Mary Balus, an attorney and unofficial legal consultant to Dr. Hurwitz, told me, "I grew up with the Constitution as my only religion. What the district attorney did to William Hurwitz scares the shit out of me. . . . Put simply, the district attorney should not be allowed to criminalize malpractice." Criminalization both raises the stakes and permits district attorneys to invoke different standards for presenting evidence and establishing cases.

Patrick Snowden was the patient who had been prescribed the large number of pills. Snowden had injured his foot so badly he required nine separate surgeries and at one point had been advised to undergo an amputation. His mother wrote Billy a letter thanking him for giving her son's life back by enabling him to deal with debilitating pain. The incident with the 1,600 pills had occurred because a pharmacy had run out of the ordinary dosage of his medication. Snowden was given two new prescriptions for pills of a lower strength, and this situation was then compounded by a clerical error. Snowden never took or intended to take all the new pills in a single day, and he was fully aware of the proper prescribed dosage. The prosecution focused on the number of pills in order to confuse the public and the jury, and they did not take into account the actual strength of the analgesics or the tolerance that normally develops during long-term treatment. The prosecutors depicted the doctor's office as being the equivalent of an old-fashioned opium den or a modern crack house filled with incoherent or sleeping people with track marks on their arms. The defense maintained that Billy was the last resort for many patients suffering from chronic pain, like Patrick Snowden.

Complicating matters, and ultimately leading to further convictions in the retrial, was the government's contention that Hurwitz prescribed narcotic medications to patients who turned out to be drug dealers or substance abusers, and therefore he was acting outside of legitimate medical practice. Jurors convicted him of drug trafficking because they believed he had ignored signs that some patients were reselling the drugs. Billy countered that he was a physician, not a police officer. He argued that it was completely unreasonable to expect him to behave like a detective when responding to suffering patients.

The jurors were required to render verdicts on multiple prescriptions provided to nineteen separate patients. This was the equivalent of combining nineteen different malpractice cases involving pain management, and then using criminal rather than civil criteria to reach a verdict. The retrial came to an end in April 2007, and Hurwitz was convicted of sixteen counts of narcotics trafficking. When John Tierney later interviewed the jurors, they turned out to be unaware that the physician might face ten or more years in prison. The jurors had hoped Billy would be sentenced at most to the two and a half years he had already served.

My friend Dr. Steve Passik, a psychologist and authority on psychopharmacology from Memorial Sloan-Kettering Cancer Center in New York, is one of the expert witnesses who testified on Billy's behalf. Steve wrote to the presiding judge:

Dear Judge Brinkema:

I write on behalf of my colleague, Dr. William Hurwitz, to add my thoughts and views as you consider his sentence. Dr. Hurwitz was a dedicated and principled physician, wholly concerned for his patients' well-being and to the cause of improving pain management in this country. Whatever his mistakes, he was dedicated to this cause first and foremost. Despite the rhetoric, he was not, as I think you appreciate, a garden-variety drug dealer. He was a smart and dedicated physician, who nevertheless made mistakes. Pain management is a complex and difficult enterprise and one in which even the most intelligent and caring of physicians can be fooled and worse. Moreover, Dr. Hurwitz practiced it at a time when historically the whole field was trying to undo the injustices of the past—to liberalize opioids [use], to listen to and believe patients. This truth bias might have played a role in his being gullible and even careless at times. I have been doing this long enough to know that there was a time before members of our field had yet to wise up to the problem of drug abuse and diversion. We used to preach that if you didn't get duped every now and again, you weren't treating pain aggressively enough. . . . I can tell you the rhetoric and the practice has changed sharply since then. What looks incredibly negligent in today's light was common practice then, and Dr. Hurwitz was determined to press that practice as far as he could to aid his patients.

Nevertheless, you will sentence him in today's pain world and your actions will be interpreted by today's pain physicians. This is a world where the doctor-patient relationship has already been dealt considerable blows by the souring of the regulatory climate. Doctors have already been pushed into adopting the law enforcement role in their practices and the result is less compassionate care and loss of trust by both parties. Rather than being therapeutic at every turn, doctors worry about the security of their practices and being viewed as drug dealers. We hardly need this at a time when the population is aging and we need more, not less, compassion toward those in pain. With your action, you can simultaneously strike a blow against doctors being viewed as criminals and for the compassionate care of people in pain.

I remember meeting Billy for the first time at a meeting of the American Pain Society many years ago. I must admit I didn't understand his willingness to take on the authorities and put himself at risk—to be a sacrificial lamb for the cause of pain management. But one thing was certain—this man knew opioids and pain management better than nearly anyone I had met at the time. He was talking to me about using ultra low dose antagonists to potentiate opioids a full ten years before I had ever heard anyone mention it! He was educated. He was dedicated. He was trusting. He wanted the best for his patients.

I hope this letter helps engender for you compassion, understanding, and a spirit of mercy for Dr. Hurwitz. I thank you for considering my input and for all of your efforts to give Dr. Hurwitz and pain management a fair day in court.

Sincerely,

Steven D. Passik, PhD

John Tierney ended his article about Billy's case by stating, "Even if Dr. Hurwitz does walk free next week, I wouldn't take much solace in his victory if I were a doctor treating pain patients. I wouldn't feel safe until doctors' prescribing practices are judged by state medical boards, as they were until the DEA and federal prosecutors started using criminal courts to regulate medicine. The members of those state medical boards don't always make the right judgment, but at least they know that there is more to their job than counting pills."

In July 2007, the judge sentenced the doctor to four years and nine months in a federal prison. Hurwitz was finally released from prison to a halfway house in D.C. and transitioned to house arrest as part of his parole and probation. He continues to live in Virginia.

Political science professor Ronald T. Libby has called for medical associations to launch a national campaign to end the unjust criminal prosecution of doctors. He has written about Dr. James Graves, a family physician who specialized in pain management and who was convicted of racketeering, drug trafficking, and manslaughter. This represented the first successful attempt by a state to apply—or, in Libby's opinion, to misapply—the Racketeer Influenced and Corrupt Organizations Act (RICO) to a doctor. The manslaughter charge alleged that he had demonstrated "culpable negligence" in prescribing medication to drug-addicted patients who killed themselves by overdosing. In 2002, a Florida jury found him guilty and a judge sentenced him to sixty-two years in prison.

A national review has reached the supposedly reassuring finding that only one in a thousand practicing doctors were charged with opioid analgesic prescribing offenses between 1998 and 2006. Approximately two-thirds of the cases entailed administrative charges involving state medical boards or the DEA, and the remaining third consisted of criminal charges. Seventy-eight percent of the physicians facing criminal charges were accused of drug trafficking or racketeering; 6 percent were being charged with murder or manslaughter.

Even less heartening was the study's acknowledgment that the data were based on charges and not investigations, as there is no question that the number of investigated physicians markedly exceeds the number charged. In addition, the review further underestimated these occurrences by limiting them to physicians and by not including nursing professionals such as Kim and Amy.

18

The Return

Amy and Kim were never officially told that the case had been closed and they could return to work. Each day Amy would call and ask to come back. Each day the administrator would say no. It went on like that for months until one day she went in for a regularly scheduled meeting with the vice president and director of nursing and the topic of the meeting was resuming work. It was the moment she and Kim had been anticipating for months, but it also brought back a flood of emotions and the uncertainty of having to be with Olga once again.

"Good old Olga," Amy told me. "Because of some legislation that protects whistle-blowers, it turned out that she had a whole lot more rights than Kim or me. She had the right to accuse us, and she had the right to return to her previous job, and the hospital had to be careful that her position was not harmed in any way. However, the problem facing the administrators was that I had no intention of working with her ever again. They figured that we would all sit down and have a nice adult discussion. I said, 'Fine, but it sounds like a terrible plan.' I just could not imagine putting us all in a room to-gether and thinking that we would have a civilized conversation—because I still wanted to punch her in the head!

"But the truth was that we would do almost anything in order to get back to work. Consequently, everyone arrived punctually at the hospital for the reconciliation meeting. It was the first time that I had seen Olga since she accused us. The nursing administrator said, 'I just want you guys to all feel free to chat.'

"I remember thinking, 'Free to chat? Give me a break!'"

Meanwhile, Olga sat silently on the other side of the table with her arms folded across her chest. Amy and Kim exchanged a glance and both looked over at Olga. The room remained entirely silent.

Finally an administrator said, "Would anyone like to begin?" His question was met by more silence. After what seemed like hours, Amy opened her mouth.

"All right, I will speak." Looking at her accuser, she said, "Olga, you will never know what you have done to Kim and me. You will never understand. You have accused us of a horrible crime that we did not commit. You have ruined our holidays. You have ruined things for our families. You have just totally ruined who we are and what we do. I guess that is really all I have to say right now."

Olga sat there with her arms folded and replied, "Well, you have no idea what it feels like to be me—knowing that you two killed somebody."

"Kim, of course, becomes hysterical and begins crying. And I say, 'This meeting is over. Come on, Kim, we are leaving.' At that point, Kim is absolutely beside herself. Kim begins sobbing and running down the hall. Finally she just falls on the floor weeping—it is awful to see her like that."

According to Kim's recollection, it was she and not Amy who confronted Olga at the meeting, and Olga replied, "Oh, yeah? Well, I will never forget how you killed that woman."

Following this comment, Kim recalled standing up and announcing, "This meeting is over!"

But Kim told me that immediately afterward she became emotionally overwhelmed. Her memory was fragmented, she explained. "I do not know how I made it to the door, but as soon as I got into the hallway I started sobbing and running. I literally forgot where I was. . . . I couldn't find my way out. Amy grabbed me, and she and the others stuffed me into a room until I could compose myself. I could not believe that anybody would think I killed a patient. . . . Given the things that I have already gone through—the physical and mental things that I endured as a child growing up—perhaps it is no surprise that Olga's words had such an impact."

Kim was still agitated when she recounted the painful confrontation. She required a couple of minutes to settle down before she was able to describe to me how she resumed working on the renal ward. She remembered feeling, "very, very, nervous about going back to work. I don't know what my colleagues are thinking. After all, the hospital is not really open or forthcoming with the staff of the renal unit about what has happened. I try to think of my fellow staff members as bystanders. I cannot expect everyone to take my side. Before this all happened, many were Olga's friends *and* my friends. Even though she has made a giant mistake, it is not fair to expect them to stop being her friends. So I am trying to get myself prepared to come back to work knowing Olga is going to be in the hospital and knowing that as much as I hate her guts, other people are not going to feel that way. Furthermore, I should not expect them to feel that way, because it really is not personal for them."

As we talked, Kim suddenly snapped back into her more ordinary state of delight and enthusiasm. She smiled broadly at me and said, "My first day, I think I am going to pee my pants—micturate in the drawers! I end up being pleasantly surprised at the way people act. I know that I have worked there for a long time, I went to school there, and so on, but the support is phenomenal! I cannot even tell you what I feel. My very best friend, Dawn, shows up at the medical center's parking lot that first morning with Dunkin' Donuts coffee and these little envelopes—twelve little miniature envelopes. She says, 'You have twelve hours to get through today. Every hour, open one of these and know that I support you.' After a big hug, I take the envelopes and my coffee, stride into the elevator, and march onto the unit.

"I am back! I go about my business, get my patients' paperwork and report sheets, and all the while everybody is saying, 'Good morning!' and 'How are you?' and 'It is nice to see you!' I listen to the report from the nurse on the previous shift. I look up, and there is Dr. Hwang holding a big box of Godiva chocolates. More hugs and well-wishes follow. By ten o'clock that morning, I have a flower delivery from a good friend of mine. About two hours later, I get one more flower delivery from another close friend. Then the vice president of nursing comes down to see how my day is going—making sure that I am feeling comfortable. A bunch of physicians arrive to express their support. People make it very, very easy to return to work. Furthermore, I open an envelope every hour and read Dawn's little cards. It is really kind of neat!"

On the other hand, Amy was considerably more somber when she recalled the resumption of her duties. "Administration decides that it is obviously not a very good plan to have the three of us return to work on the same floor—no kidding. They move Olga to another ward, and Kim and I return to the renal and transplantation unit.

"We are all back at work. But those first few times taking morphine out of the machine, I have to say I am a little worried somebody is watching or that I will do something wrong. And I start questioning every single thing about myself and about nursing. That is when I decide to step down from our clinical ladder—I drop the administrative part of my job. I just want to be a basic nurse. When the medical center's chief executive officer finds out about this, he tells me the clinical ladder is meant to be climbed. I say, 'Well, my understanding is that if it is a ladder, one can climb up or down, and I am choosing to climb down.'

"I climb down and am staying where I am. Less money, prestige, and esteem go with this decision, but I am satisfied. It certainly puts things in a different light when you have the state police arrive at your door.

"A few more weeks go by and I realize that no one has ever officially informed us that the case is over. I make an appointment with the CEO, tell him what it is like to have an open murder investigation hanging over my head, and ask if there is any way he can tell me the case's status. He picks up the telephone and calls a hospital attorney. After a brief conversation he looks at me and says, 'It is closed. It is all done.'"

Kim apparently never learned about this conversation between Amy and the CEO; alternatively, she correctly recognized that there actually is no statute of limitations for murder investigations. During one of our talks, she exclaimed, "Meanwhile, I understand that we have gone from being a hot case to one that is sitting on a shelf. The district attorney cannot find anything nefarious, so our case has become the one that they will get to when they get around to it. I do not know if they will ever get to it. They are supposed to inform us, but we never hear anything official from anybody."

I was unsuccessful at my attempts to speak with the district attorney, William Bennett, and eventually asked a lawyer acquaintance to help me clarify its status. A couple of days later, I received an e-mail stating, "The DA's office was somewhat cryptic and would say only that the investigation is inactive and they do not anticipate actively pursuing it in the future."

The e-mail continued, "Reading between the lines, I conclude that the DA found nothing, and will find nothing, and that the investigation is for all intents and purposes concluded. Perhaps it is only ego preventing him from saying so."

Well, perhaps it is ego, but I think it is more realistic to acknowledge that the reason cases like these are never completely closed is a prudent one—in the event that Kim and Amy are ever again accused of misdeeds, the record of this investigation will be readily accessible.

Olga was distressed because from her perspective the case had been swept under the rug. She came back to work on a general medical ward at Baystate rather than the renal unit. She resumed her friendships with people at the medical center and tried to get over her disappointment with the district attorney's office and the hospital. During one of our telephone calls, I told her that I appreciated the courage it must have taken to step forward and to become a whistleblower. However, she did not take this as a compliment. Instead, she described how on one occasion she went to change her clothing at the conclusion of a shift and discovered that someone had taped a whistle to her locker. She felt taunted and victimized.

Olga's work experience continued to be acrimonious. I learned from an administrator that Olga subsequently claimed a fellow TA on her new unit was responsible for another patient's death. Months later when I approached the accused TA, she was still too upset to talk on the record about what happened. From what I could gather, Olga thought she was suctioning the mouth of a comatose patient too vigorously and that this hastened the person's demise. Following an internal investigation at the hospital, the staff member was exonerated.

Given this latest occurrence and the murder accusation of Kim and Amy, it seems as though Olga views the world as a chaotic place inhabited by impulsive individuals whose behaviors can be unethical and illegal. She clearly feels compelled to actively confront such situations and to draw the attention of authorities toward investigating and correcting them. Olga also relies on external strictures—laws, punishments, and so forth—to maintain a feeling of safety. Many individuals who embrace the certitude offered by organized religion with its reassuring and clearly defined precepts on how to live can take comfort in the rule-and-regimen-bound environment of hospitals. Radically different from that of Kim and Amy, Olga's moral framework was threatened by the other nurses' willingness to cross boundaries to relieve suffering.

Ultimately, Olga was not interested in making exceptions. Although I believe that the conflict between the three Baystate staffers is a symptom of America's unease with the philosophy of palliative medicine, on another level I think that it resulted from a collision of not only beliefs but also personalities. Unfortunately, Olga has found that stepping forward and pointing a finger can precipitate tragedy.

During a casual chat with Amy at the renal unit's nursing station, she began to chuckle and described accidentally encountering Olga Vasquez. "I was walking out of the employee pharmacy," she said, "and we literally, boom"—Amy clapped her hands—"walked into each other.

"At that moment, I do not even realize who I am apologizing to as I say, 'I'm so sorry, I just walked out of the pharmacy with my eyes closed.'

"Olga automatically says, 'I'm so sorry.' Then she realizes it is me. She becomes very animated and exclaims, 'Oh, my God, I cannot believe it is you! Oh, my God! I am so sorry, so sorry.' She is profusely apologetic.

"We don't have much of a conversation because it is lunchtime and there are probably 3 million people in the hallway. I have not seen her for a while. She has cut her hair. She looks great. I tell her, 'You look beautiful.'

"She says, 'I am so sorry for what I did,' She catches me completely off guard. I think her exact words are, 'Can you ever forgive me for what I did to you?' and I reply, 'Yeah, I can forgive you for what you did.'"

I cut in at this point and inquired, "Can you?"

Amy replied, "Absolutely. . . . It still does not mean that I don't want to backhand her. But I can forgive her." She let out a loud guffaw. "I can knock her down and kick her, but I can also forgive her." More guffaws.

"It does not do any good to hold a grudge forever, and I do not know what was in her head. I really don't. It was not like her. She was a friendly, churchgoing, God-fearing, born-again Christian. She had values and morals, and she was a kind person. She had a big heart and took very good care of her patients. I never heard a complaint about her or from her—never, never, never. The accusation came out of left field, and it was a totally freaky thing."

During a subsequent conversation with Olga, I asked her about the run-in with Amy. She explained that it had been a couple of years since they'd last spoken. She remembered Amy embracing her in the corridor and that she hugged her back and cried. In contrast with Amy's account, Olga denied retracting her position or apologizing. She was sorry, but not about anything she had thought, said, or done concerning the murder accusation.

19

A View from the Bench

The majority of accusation cases involving innocent nurses and doctors take place because of dissension within the clinical team; families are involved only secondarily. The accusers are often staff members who are not likely to have had a voice in clinical decisions. They may be positioned on a lower rung of the medical hierarchy, and they are unfamiliar, uncomfortable, or unaccepting of the current palliative medicine ethos. Murder accusations are derived from a thought process that is two parts emotional to one part cognitive—they represent a clash of differing beliefs and of personalities. But they may also be partially triggered by the recognition of a real or perceived medical error that appears to the accuser to have been intentional, such as Kim's insistence that the oxygen be removed, or by the perception that the clinician is deviating from medical guidelines.

The most effective form of legal defense is not to have cases like these filed in the first place. Jack Schwartz, an assistant attorney general for the state of Maryland, told me during a meeting at his

Baltimore office, "Keep the nurse's aide from going to a lawyer by educating her as to why it is okay to stop dialysis and why it is okay to pull a feeding tube. . . . Even if the nurse's aide has religious beliefs and would not want anyone to withdraw life support if she herself was dying, there needs to be an intellectual overlay that keeps the emotional reaction from translating into her taking the first unfortunate step that begins the legal chain of events."

Schwartz' point is that there needs to be a better understanding of how health care decision making works, and it must be clarified that patients or proxies have a right to request, defer, or refuse treatment. His solution calls for educational efforts that involve everyone who has contact with the patient—beginning with the person who sweeps the floor and cleans the bathroom. Educational outreach is needed to communicate just how complicated it is to exercise those rights—it requires remarkable fortitude to arrive at a decision to stop a life-prolonging treatment—and how any resulting death is not the equivalent of clinical suicide or murder.

However, one can legitimately question whether education is the sole solution or whether something else is needed. I think people who disagree with the palliative care ethos already understand how hard it is to come to the decision to end a life, but they don't consider it to be a morally acceptable decision under any circumstances— and explanations will not change their opinion. It seems that a more nuanced policy determination about the exercise of individual conscience in medical practice will be necessary at some point. Perhaps didactic efforts need to be supplemented by enlisting representatives of the lower echelons of the medical hierarchy in the decision-making process.

Jack Schwartz was smirking, but he was also quite serious when he said, "At the same time, the attorney generals need to attend to the rare crazy doctors, crazy nurses, or deeply misguided health care professionals who break the law and murder patients. Law enforcement cannot say—like in Vegas—whatever happens in the hospital stays in the hospital. Instead, there will always be a need to occasionally investigate. And if a complaint comes in, there is a need to have a basic understanding about what the ethics are—because the law in many states, like Maryland, tracks the ethics. The law is built upon ethical concepts."

Attorney generals are the chief law enforcement agents of the states, and they can play a valuable role in articulating why the law is not a barrier to stopping life support or providing analgesia when staff documents the appropriate medical steps. If the AGs are unfamiliar with current medical practice, then they need to consult their more knowledgeable colleagues, such as Jack Schwartz, and they also need to form closer working relationships with medical professional boards. As part of the investigatory process, consultations with palliative care specialists are now legally required in Arizona; this policy should be adopted nationally.

Rosie's case took place in bucolic western Massachusetts, but one can easily imagine how a major catastrophic event might tear apart a community's social network and lead to similar allegations. Following Hurricane Katrina, murder accusations were leveled at Dr. Anna Pou and two nurses at Memorial Hospital in New Orleans, and that experience has led Louisiana to pass a law redefining the role of physicians and other health care professionals in response to disasters. The legislative reform authorizes the use of medical panels to review the clinical circumstances of cases being considered for criminal prosecution by the district attorney or attorney general, provides additional liability protections to medical personnel, and calls for new standard-of-care guidelines applicable upon declaration of disaster by the president or governor.

Louisiana, like other states, already had a Good Samaritan statute that provided limited protection for health care professionals. A doctor who arrived on the scene of an accident and rendered gratuitous services was protected against litigation for negligent conduct. The new laws can be considered an expansion of the Good Samaritan statute to encourage medical personnel to respond to disasters without risking litigation.

Another way to prophylactically prevent cases from becoming criminal matters is for medical facilities to take greater advantage of their own internal systems for reporting and examining complaints about care. For example, Baystate has established multiple means for its staff to express concerns about patient welfare—the hospital sincerely wants to learn about every possible error of practice or judgment. Staff can directly approach managers, leave voice messages at a special number, or send e-mails to a particular address. Complainants can identify themselves or choose to remain anonymous. Every medical facility should have similar warning and notification mechanisms, along with the necessary committees to explore staff concerns and take corrective actions. Modern medical centers have a tradition of tracking medication use and patient deaths; when combined with staff complaints, the association between these factors has led in several cases to the arrest of bona fide serial killers. Whistle-blowers need to be appreciated for the courage it takes to step forward and try to correct perceived mistakes or decipher complex issues.

Whether cases are handled internally or externally, the investigations should be conducted with as little drama as possible. Just because there has been a complaint, one does not want to irrevocably disrupt the lives of people who are earnestly trying to provide palliative care. While I was initially miffed that I had to learn about the Baystate case accidentally and long after its conclusion, I now admire the medical center for having so effectively maintained confidentiality.

Upon investigation, even if a clinical error is found, there are steps that can be taken that do not entail overly harsh and punitive measures. There are clinical protocols that may simply need to be adjusted or professional licensing boards that can be consulted. In all of these cases, one wants to try to avoid an overly rapid law enforcement response for an event that may turn out to be either a misunderstanding or a shortfall in quality care—but not a criminal act.

In my desire to further elicit a law enforcement perspective and to discover whether Kim and Amy responded appropriately, I decided to meet with a judge. The security personnel who scanned me and my possessions, as well as the police officers armed with automatic weapons who were noticeably patrolling the outside of the courthouse, all looked as if they had previously been linebackers in the National Football League. They were certainly not the relatively diminutive and friendly folk one sees manning security lines at airports. My tape recorder looked suspicious to them and immediately was taken away (and never reappeared).

Before I actually sat down with U.S. District Court judge Kathryn Basham (not her true name), I was already thoroughly impressed by the security and the dangers that are faced by someone in her position. I had known that judges in Italy or Latin America were at risk of being assassinated, but was unaware that the dangers are just as real in the United States. Basham had presided at the trial of a health care worker who was found guilty of multiple homicides and received a life sentence, and I sought her out because I figured she would be well acquainted with the issues surrounding accusations of murder involving medical personnel. Although I certainly had no nefarious intentions, by the time the judge and I sat down together my heart was beating rapidly. Kim and Amy had told me about their terror at being suspects, and I probably experienced the tiniest scintilla of this in my visit to the courthouse.

Basham understood that many accused health care professionals lack any malevolent motives, and her recommendations were obviously sympathetic and heartfelt.

"If my next-door neighbor is a nurse," she told me, "and she knocks on my door seeking advice about having just been accused of murdering a patient, the first thing that I would do is sit her down in my living room and give her a stiff drink. Then I would explain that she needs to 'lawyer up.' She must immediately seek out the best homicide attorney in the state and hire that lawyer to represent her in the proceedings. It is going to be expensive, but it does not matter if she is going to have to refinance her house—the potential penalties in a murder case warrant any expense.

"Her professional career, relationships with loved ones, freedom, and life are all threatened. She has to avoid the temptation of securing the legal services of someone who is inexpensive. My neighbor definitely should not accept the kindly offer of assistance from the congenial attorney who drafted her will. What she needs is the active attention of the most successful and experienced homicide lawyer who is familiar with her state's legal system. While it is completely unfair, she is not going to emerge from this ordeal as the same person who began it.

"She also needs to refrain from saying another word to the police investigators—they may appear to be genuinely interested in clearing up this 'misunderstanding,' but she has to understand that their motives and goals are not necessarily the same as hers. To 'lawyer up' basically means to shut up. Nurses and health care professionals are accustomed to reflexively helping others, but the only person who realistically needs help is herself. She needs to rely on the advice of a seasoned criminal attorney in obtaining it."

Basham went on to explain that absolutely everyone involved in law enforcement sees the world through jaundiced eyes. The same detective or district attorney who offers to assist in untangling the allegation has already seen far too many corpses and crime victims to believe in innocence. Their worldview is colored by a not-so-subtle belief that everyone is guilty of something, and their role is to identify and punish miscreants. It is perhaps not a surprise that the combination of an accusation and the discovery of even the slightest medical misstep—such as not having narcotic wastage witnessed by a colleague—counts as three strikes in their eyes. Basham's account was enlightening, and I began to understand more fully why the Utah district attorney's office overstepped itself in Robert Weitzel's case, and why assistant attorney general Charlene Malone and the other prosecutors continue to this day to believe he got away with murder.

During my separate interviews, I found both Basham and Schwartz to be deeply thoughtful and utterly engaging. Each frequently punctuated observations with wry, intelligent wit. I have grown accustomed to the mocking humor that allows medical staff to function amidst the ugliness of pain, suffering, and bodily deterioration, and the amused comments of these law enforcement representatives appear to have come from a similar place. I was forced to conclude that although they may not feel lucky, Kim and Amy were incredibly fortunate the grand jury did not indict them. They really could have ended up in the Big House.

A decade ago, an attorney named Ann Alpers wrote a prescient article in the *Journal of Law and Medical Ethics*: "Criminal Act or Palliative Care? Prosecutions Involving the Care of the Dying." Alpers documented over an eight-year period at least twenty-three investigations of professional caregivers, eight indictments, four murder trials, and two physician convictions. As only ten physicians between 1935 and 1990 had been previously charged with killing terminally ill patients in the United States and none of those had served any jail time, she was forced to conclude that this represented a substantial increase in prosecutions.

Professor Stephen Ziegler of Indiana University–Purdue University in Fort Wayne has expanded on this work and specifically has been interested in examining prescription drug prosecutions, I approached Ziegler after reading several of his articles about prosecuting attorneys' views on pursuing cases related to euthanasia or aggressive use of opiates. Stephen had previously worked as a police officer and detective for ten years in Hurst, Texas. He had careers as an assistant prosecutor, a defense attorney, and a task force officer for the DEA before becoming an academic.

Stephen maintains that many of the decisions by state prosecutors to investigate, refer, or indict are not prompted by overzealousness, but most likely stem from a lack of knowledge concerning prescribing and related practices. He sees an opportunity for state medical boards to take a leadership role and become a reliable resource and "clearinghouse" to both the legal and medical communities. He thinks a comprehensive approach is necessary to deal with the prosecution of physicians and nurses for end-of-life matters. He suggests that drug and medical regulators—broadly defined to include local prosecutors, local police, and the Drug Enforcement Administration—each has the right to investigate physicians or nurses, but would greatly benefit by having a place to turn to for answers to these complex questions. An ad hoc committee ought to be created within each state medical board to offer expert suggestions and advice as these situations arise. The committee should include members who are knowledgeable about modern pain management and palliative medicine as well as national and local bioethical and medical standards.

The dialysis clinics in our area hold annual services of remembrance to honor patients who have died in the previous twelve months. The services vary according to the clinic, but each extends invitations to families and staff for an opportunity to meet once again and recall the multiyear periods of treatment and the circumstances of each death. The services sometimes occur in the largest hospital auditorium—

which they fill to capacity—and other times at local churches. In either venue, they are designed to be ecumenical and include inspirational readings, poems, and musical selections from diverse faiths and ethnicities. There is always a corkboard that families use to display photographs of their deceased loved ones, and a candle-lighting ceremony during which each patient's name is spoken. There is an opportunity at the conclusion for families to break bread with staff, recall poignant memories, and offer thanks.

I try to regularly attend the services of remembrance because they allow me a few necessary moments for reflection. Nowadays, as I've mentioned, four out of ten deaths at New England clinics are preceded by decisions on the part of patients, family, or staff to stop dialysis. Sometimes I think that those individuals were especially heroic. Other times I think that the ones who kept up the treatment to the bitter end were the heroes. Looking at the hundreds of people who solemnly assemble, it is impossible to differentiate between the types of deaths or to determine whether there is any differential impact on the bereaved loved ones. I remain personally pleased that discontinuation is an option.

During the past seven years I have read and reread Kim, Amy, and Olga's words and have come to greatly respect and admire each of them. The stories of the individual patient deaths and the accounts of the Baystate staff inspired me to begin a journey to make sense not only of their individual conflict but also of what I discovered to be similar accusations that are taking place across the country—and, for that matter, around the world. I have found that nurses and doctors are increasingly accusing each other of murdering their patients. Most of the disagreements are handled quietly within medical facilities by Solomonic administrators. Sometimes they are dealt with by hospital ethics committees or state boards of registration. On occasion they are brought to the attention of law enforcement representatives and then become criminal matters. In all of these cases, accused and accusers are caught in the middle of society's profound confusion and ambivalence about end-of-life issues and the choices of terminally ill patients to have treatment withheld or withdrawn.

For two decades, medical ethicists have maintained that there is a moral equivalence between withholding and withdrawing life-prolonging treatments. In the United States, this has been paralleled by consistent decisions by the courts that find these practices to be equally legal. However, the Terri Schiavo case and the coalition that it has reinvigorated now dispute these claims. Echoing the beliefs of the more conservative elements of Catholicism and some of those held by Orthodox Judaism, they question the ascendancy of the ethical principle of autonomy, the right of people to judge quality of life, the reliance on advance directives, and whether withholding and withdrawal are morally and legally equivalent. They claim that the discontinuation of treatment is fraught with difficulties, and they assert that artificial nutrition and hydration should never be fully stopped.

An international perspective shows that these same issues are being actively debated and that there are distinct differences in the practice of medicine between the countries of southern and northern Europe, the Middle East, and Asia, and between First World and Third World nations. The global challenge faced by modern medical clinicians is the desire to accommodate the very different requests of dying patients. On one hand, medicine would like to respond to the demands of the terminally ill writer Susan Sontag, who made it completely clear that she wanted absolutely everything possible done for her recurrent metastatic cancer. On the other hand, medicine has no desire to force-feed life upon the unwilling—such as Winston Churchill, who after age and decay had taken their toll remarked that in the previous five years "I have had no appetite for life. No—when it comes to dying I shall not complain. I shall not miaow." The

Pulitzer Prize–winning columnist Art Buchwald arrived at a similar point in the course of his diabetes, peripheral vascular disease, and chronic kidney disease, when he, too, decided that "enough is enough." The current practice of medicine permitted him to discontinue dialysis. While being visited at his residential hospice bedside by a parade of friends, family members, and celebrities, Buchwald characteristically quipped, "Dying is easy. Parking is impossible."

My chief hope is that the public discourse will quickly move beyond emotionally based arguments and murder accusations. Instead, a meaningful dialogue that produces thoughtful and balanced policies is desperately needed. In the end, both palliative care practitioners and members of the coalition desire compassionate end-of-life practices that respectfully and flexibly navigate the differing beliefs of patients, families, and staff. Accomplishing this complex task will be a fitting legacy not only for Terri Schiavo but also for Barbara Dilanian, Benjamin Babcock, and Rosemarie Doherty.

20

Moving On

In 2005, I arranged another formal, restaurant-based interview with Amy. I saw her outside a cafe in Northampton, and the sight broke my heart. She was jauntily wearing a white baseball cap to conceal hair loss from her latest round of chemotherapy. Amy smiled and lent me a quarter for the parking meter. She was twenty pounds thinner, and later I would watch at the restaurant as she consumed a total of three forkfuls of what appeared to otherwise be an excellent meal. She had been diagnosed with stage II ductal infiltrative breast cancer, had undergone a mastectomy, and was planning on having a second one performed prophylactically. I had known that her grandmother and mother had died of breast cancer but had been unaware that she had also lost two aunts to the disease. For years Amy attended a high-risk breast program, but she was only belatedly diagnosed with the cancer.

"I'm ready today! Let's book the operating room!" Amy announced at the first appointment with her surgeon.

Instead, she had to wait ten days—"which," she explained, "was an eternity! I know it was comparatively fast and some people wait for weeks and weeks and weeks. It may have been relatively quick, but they were the worst days I have ever lived through . . . except for the little thing with Olga Vasquez." She laughed nervously.

Perhaps I could or should have steered our luncheon conversation onto more pleasant topics, but much of it revolved around her mother's final illness. Amy's devotion to dying patients and her need to communicate with them is clearly based on this experience. She explained, "At the time, I was just a stupid kid. I was only sixteen. Now I would liked to have sat down and talked to my mother about what it is like to die. . . . Because that is what I always try to find out from my patients—what it feels like when they are dying.

"During the last five or six months [of my mother's life] she was total care . . . sick, sick, sick. But we never talked about dying. Back then you didn't have conversations about dying. She had metastases to her liver and her skin was completely yellow with jaundice. She probably weighed fifty pounds soaking wet, and all of it"—Amy extended her hand outward from her abdomen—"was liver."

During our lunch, Amy and I conversed again about Olga. She concluded, "I don't hold a grudge now. I absolutely forgive her. But I would love to know what was in her head. I really would love to know. What she did was so outrageous. There was no friction leading up to it—we were so friendly! It is all very strange."

In 2008, I met with Amy again. This time she looked vigorous and strong. There were yet another couple of years to go before she would reach the magical five-year point of being cancer free, but she seemed quite healthy and was proud at having recently run and completed two 10K races.

We talked about religion, and she alluded to having been raised as a Catholic. "We did it because we were told to do it . . . first communion, confirmation, all of that. And now I don't consider myself as belonging to a religion anymore. It's not my bag. I don't really have a bag right now. If you're a good person, you treat everybody equally well. My mother used to say, 'Do unto others . . .' I would like to think that is pretty much how I live my life. I don't worship anything in particular." Amy and I chuckled together as she amended her last statement: "Well, during the summer I worship the golf god."

Soon Amy slipped into a reflective mood. "I think back to dealing with people at the hospital and having conversations about the end of life. Whatever you believe in is fine, if you are comfortable with it.

"I remember one woman in particular who was dying of cancer and who told me her vision of life after death She pictured family and friends—everyone who had a special meaning to her—forming a big circle in which they had their arms around each other. She believed that when you died one stepped back out of the circle and the remaining people then reconnected. She was thinking about being on the outside and looking in. . . . When the woman recounted this to me, I thought that it was quite a neat way of looking at death."

Amy explained that her mother had a similarly sanguine view of Heaven—a place suspended on top of the clouds where everyone languidly reclined in chaise longues, hung out with buddies, and got the tans they had always wanted. While Amy was not necessarily prepared to believe in either of these concepts, she also has no compunction in confiding them to her terminally ill patients who were seeking answers or solace in the face of their own existential dilemmas.

Amy looked me straight in the eye and remarked, "After all, what do you say to people who are grappling with imminent death?" And once again, I appreciated that this nurse spent her day listening and trying to comfort dying people while I and other physicians stood to the side and occupied ourselves with other matters.

Amy brought along the following letter concerning Barbara Dilanian's death:

Dear Amy,

I have thought about you and the staff at Baystate often, but I was not ready to write this until now.

Thank you for allowing my mom and my family to experience some final moments of happiness. The wedding that all of the wonderful people at Baystate Medical Center gave our family changed our lives. You took a tragic time and changed it into a memory we will always cherish.

Allen and I have celebrated our first anniversary and are almost up to the second! I truly feel I have had an easier time coping with my mother's death, because of the amazing gift you gave us. Whenever I get cynical or feel the world is a cold place, I think about the empathetic, caring people I met at Baystate and their selfless act. . . . I feel lucky to have met you all.

> *Sincerely,*
> *Jane Reitz*

In 2008, I spoke with Jane and her brother, Ken Dilanian. He had become a reporter for *USA Today*, was married, and had a child of his own. He described Jane's wedding on Cape Cod as having been a wonderful event that took place at a lovely church in Chatham. Jane now had two children and was still happily married to Allen. She told me that she considered herself to be a spiritual rather than religious person, and she believed there was a special connection between her firstborn child and her deceased mother. Barbara's final illness had left Jane with the conviction that families need to act as strong patient advocates. This especially included insisting on the preservation of quality of life at the end of life.

Kim was able to successfully return to her nursing shift on the renal and transplantation unit. She continues to be highly regarded by her peers and has come to feel more secure about her position at the medical center. She decided to continue her nursing education, and applied for and was awarded a coveted scholarship in a master's-level nursing program. I discussed her success with Eileen Grunwald, the hospital administrator, who was thrilled at the thought of Kim eventually becoming a senior nursing instructor at Baystate.

During my final interview with Kim, shortly before Thanksgiving 2008, she was ecstatic at being one semester away from graduating. Only one other person in Kim's extended family had attended college, and she will be the first to complete a master's degree. As she sat across the table grinning, I realized this was the only time I had ever seen her with unbound hair—it was usually wrapped in a tight bun or a ponytail so as not to interfere with her hands-on care of patients.

At this last meeting, Kim told me about the recent death of her grandmother, who was chronically ill with diabetes and emphysema. Kim and her grandfather had been summoned to the intensive care unit, where they were informed by an attending physician and a young house officer that the patient was refusing to be intubated and placed on a ventilator. Her grandfather immediately protested, and Kim told him, "Whoa, Grandpa! You don't have to like her decision. But she gets to make it. She should decide what she wants to do."

Kim said to me, "And it was funny to glance at the doctors watching me tell him what we had to do as a family. . . . But as soon as we got into the room my grandfather broke down anyway and said, 'Oh, Mother, you have got to try! Just don't give up yet! You have got to try at least one time!'"

Frustrated at his reaction, Kim took his hand, then turned to her grandmother. "Grandpa loves you, and I love you," she told the dying woman. "He doesn't want to see you die. But you have to understand that the choice you make right now, you have to live with it. If they put this tube in, it may never come out. You may never be able to breathe again without the help of the machine as long as you live. That's what this decision might lead to."

Kim's grandmother looked imploringly at her elderly, distraught spouse and said, "I just don't want to do it. Don't make me do it."

When he saw her face he composed himself and said, "Well, we'll do whatever you want."

Over the next hour, Kim's grandmother received some morphine. Her room was filled by a succession of visiting family. Several times she singled out Kim and said, "Will you help me to die?"

Kim understood this to mean that her grandmother had listened to her stories about work and knew Kim would do everything to ensure her comfort. Her grandmother trusted her and was calmed by her presence.

More and more family and loved ones dropped in to say good-bye, including Dawn—Kim's best friend, who had provided the little cards to help her get through the first day back at Baystate. Kim's grandmother was accustomed to regular visits from Dawn, and lit into her for not having appeared during the previous three weeks. From her deathbed, the woman lovingly exclaimed, "Where the hell have you been?"

Kim's bright blue eyes blazed as she proudly held up her hand and showed me her grandmother's sparkling wedding ring. "My grandmother believed in God, and believed that when God was ready to take you, he took you. She also recognized—perhaps from conversations we had together—that medicine is not God. She wanted me there, and she knew that I would do everything within my power to comfort and ease her suffering."

I tried calling Olga repeatedly but had a hard time reaching her. Sometimes I left a message; other times I did not. When I finally got hold of her, Olga was all warmth and apologies for not phoning back in response to my messages. She quickly brought me up to date with her life.

The conversation was much like our previous interviews: my sense was that despite being surrounded by family and other people, Olga was a lonely woman who urgently wanted to connect and to relate her story. This feeling was reinforced by her account of having been a latchkey child who would return home from school with her siblings and await the eventual arrival of their hardworking mother. She spoke minimally about her father but left me with the impression that his job in the merchant marines had led him to be absent from their family for weeks and months at a time.

Olga was dissatisfied with the low status of a TA, and she had loftier professional aspirations. After graduation, she left Baystate for a more responsible position at a skilled nursing facility. I was pleased to learn she had continued her education and was now a licensed practical nurse. She was both proud and optimistic that her educational accomplishment presented a positive example for her children.

At the nursing home, Olga cared for a succession of dying patients. She was emphatic about how hard she tried to attend to all of their needs. She was repeatedly disappointed with the patient families and how they were rarely present at the terminal moment. This same theme was echoed in her recollection of Rosie's final hospitalization, and I understood for the first time that much of her indignation was connected with the belief that the patient's son had not been able to be present at his mother's death because the dying process was artificially accelerated. It was left to Olga to be the sole witness to Rosie's death. I had not sufficiently appreciated that she and Kim shared a strong desire for terminally ill individuals to be attended by their loved ones as they expired. But whereas Kim and Amy did not hesitate to alleviate suffering by taking steps that would also shorten their patients' lives, such as discontinuing dialysis or liberally administering medications, Olga found these practices to be intolerable. Furthermore, in her capacity as a TA at Baystate, she could not give orders, and instead was forced to passively comfort patients and endure their death throes. Olga may have been correct that Rosie could have received some benefit from oxygen—or at least from wearing a breathing mask—but she was not empowered to effectively provide this. She had been absolutely horrified when she heard Kim tell Rosie that it was okay for her to die, and this fueled her other concerns.

Our conversation concluded with my recognition that Olga had always done her utmost to maintain her marriage, raise her two children, care for her patients, and improve all of their lives. I learned that in the past couple of weeks, she had left the nursing home and had started at a better position in a community hospital. Despite exhaustion from regularly awakening at four-thirty or five in the morning to arrive at her job for seven o'clock rounds ("Not seven-oh-one, but seven," she clarified), Olga still had dreams. In 2010, she intended to begin taking the necessary courses to become a registered nurse.

Within a few days of this final conversation, I received several voice messages on behalf of Olga. Her family members were still concerned about how she would be portrayed in the book, and they wanted me to use a pseudonym. I realized that they could not be sufficiently reassured, so I acquiesced to the request. In the end, only two names in this book have been altered—that of Olga and the federal judge (the latter wanted to be unreservedly frank in the interview and did not wish to deal with any ramifications from her statements). Everyone else granted me permission to recount their stories and use their real names.

I was never able to reach Benjamin Babcock's daughter, grandson, or aunt. Doris Costello, a Springfield neighbor who provided me with charming anecdotes about Ben, proceeded to call six of his closest acquaintances in her search for a family member. Although unsuccessful, she and the other friends unanimously agreed that I should use Ben's true name and pay tribute to this remarkable man.

Jim Doherty was Rosemarie Doherty's eldest son, and I called him after realizing that I knew almost nothing about his mother other than the bare facts of her medical care. He filled in some of those details, beginning with her long-standing health problems that included high blood pressure, chronic dialysis maintenance for renal failure, and emphysema from "smoking like a fiend." She was a very active individual until fracturing her hip, two or three ribs, and a collarbone in an automobile accident. Despite comprehensive treatment and a stay at a rehabilitation facility, she never really re-covered.

Before the accident, Rosemarie had been a vibrant and stubborn soul who had lived through the Great Depression, worked hard her entire life, cared for her parents, raised three sons, and regularly glowed in the company of three grandchildren. According to Jim, "My mother's door was always open. A neighbor having problems at home with a husband, kids, or just the grind of life could always find the absolute worst cup of coffee in the universe—on this I am not kid-ding, as Rose was not known for her cooking acumen—and an open ear. And believe it or not, folks came back again and again—and not for the coffee. She loved to laugh and wasn't hard to please. She took absolute joy in the simple act of walking. She walked the streets of Holyoke for years—on the coldest days of winter and the hottest days of summer she would still walk five miles. Rose was always ready to help. She was a devoted and loving mother and grandmother. Rose was a proud woman, ever courteous, always with a 'thank you' and a 'please,' never rich, and never afraid to share. That she suffered is true, that she never uttered a complaint is also true. She was strong and brave. It was a privilege to have Rosemarie as a mother."

Jim wanted me to communicate to Kim and Amy that his family was thoroughly impressed with the superb care they provided. He recalled telling the police detectives that the treatment his mother received at Baystate was "phenomenal." Jim also explained to me that he was the one who had originally raised the possibility of stopping dialysis when it was obvious Rose was not improving. The physician readily agreed with his assessment and the cessation of dialysis had the full backing of the entire family.

Directly before Rose's church service, Jim had received a call from a friend who ran the local funeral home. The friend notified him that the police wanted to "impound my mom's body." The two men decided instead to get the corpse to the church, where everyone was waiting. Following the service, Jim met with a Lieutenant Higgins and one of the assistant district attorneys.

The DA asked, "What would you say if I told you that someone gave your mother too much morphine?"

Jim responded, "To be perfectly honest with you, I would like to know who it was so I could say thank you."

He went on to explain to me that he had fully understood his mother was not going to get better, and if he had known someone was going to give her a little more medication to make her comfortable, he would have thought that was a good thing, not a bad thing. He also told me, "If you knew my mother's sense of humor, [the police's rush to claim her body] would have been her last laugh—it was a riot!"

Jim released the body on behalf of the family, and the medical examiner took possession and performed his investigation. When the toxicology did not support the euthanasia accusation, Mrs. Rosemarie Doherty was cremated as originally planned. Her ashes are still in Jim's closet.

Acknowledgments

I love acknowledgment sections and read them as assiduously and with as much pleasure as my mother-in-law reads the obituaries in the *New York Times*. I am fully aware that most people skip this part and most authors merely use it to credit the behind-the-scenes contributors—for example, "I would like to thank my incredibly talented literary agent, Harvey Snodgrass, who got me as large an advance as was humanly possible." However, I regularly and carefully sift acknowledgment sections for the pleasure of discovering autobiographical and other details. To a greater or lesser extent, acknowledgment sections reveal the man or woman behind the curtain—the Wizard of Oz.

"You are God's favorite child" was what my mother used to whisper when I was quite young. Because she was my mother I didn't doubt her words, but with the completion of *No Good Deed* they have taken on a new meaning. This book has been an exceedingly intense and personal odyssey, and I do not think that I could have completed it while simultaneously balancing my responsibilities as a physician, husband, and father without being the recipient of Divine favor.

The journey also took a toll by requiring me to become

uncharacteristically expressive and self-revealing. By nature and by profession, I am a listener and not a talker. I am basically timid and shy, and until beginning *No Good Deed*, I lacked any desire whatsoever to address a general audience. Even in writing this book, I would have been entirely satisfied to remain entirely in the background as an omniscient narrator, but there is something about the nature of accusations of murder aimed at health care professionals that forced me to speak out publicly. Although I have never been subjected to any of these allegations, my unusual behavior is clearly related to my own hitherto private experiences and beliefs about death.

One of these experiences occurred in infancy: I nearly died. When I was a year old I suddenly developed an intussusception—a condition in which the bowel telescopes in on itself and becomes acutely obstructed. According to family lore, I was cured by a simple but novel surgical intervention that would have been unavailable if I had not been born in New York City at that particular moment in time. I had the good fortune to be cared for by a singularly dedicated and innovative surgeon, and the heroic physician literally saved my life.

Aside from my choice of medicine as a profession, the event had a number of other consequences. For example, in my childhood I never once daydreamed about what it would be like to live during the Golden Age of Greece or in Victorian England, because I would have died during infancy. More important, I have always been exquisitely aware of my own mortality and the fragility of life, and I have never been especially intimidated by thoughts of dying. Although Sigmund Freud, the father of modern psychiatry, maintained we cannot grasp the immensity of death in our unconscious (which is why we never die in dreams), I routinely live with a vivid awareness of how easy it is to perish. I greatly enjoy being alive—but I know that a time will inevitably come for me to expire, and although my family will mourn my demise, I am simultaneously comforted by the knowledge the world will continue to spin on its axis.

Accordingly, decisions by patients and family members to accept and to even embrace death have always made sense to me. People whose lives are artificially extended through the use of medical treatments and whose deaths can be precipitated by stopping those treatments appear to me to be extremely fortunate. I take pleasure in their having been given the gift of additional time—as I was—and I sometimes envy the freedom to depart this world in a potentially graceful and controlled manner.

When my mother was found to have a glioblastoma—the same type of brain tumor from which Senator Edward Kennedy died—I was neither surprised nor displeased when she stoically accepted the neurologist's assessment that there was nothing meaningful that could be done to alter her fate. Without the slightest *Sturm* or *Drang* she resisted any temptation to engage in futile treatment and was dead within a matter of weeks. I had no qualms about her decision.

Likewise, when my elderly father developed a urinary tract infection, my family and I were eager for him to undergo a course of oral antibiotics. But he was severely demented and had not spoken a word in years, so when the pills proved ineffective, we were emotionally and cognitively prepared to oppose the recommendation of his primary care physician to admit him to a hospital. His doctor wanted to administer more potent intravenous medications, and he could never comprehend why we would allow the sepsis to continue and for our father to die at home (this was before hospice services became widely available in the United States). Again, it just seemed like the natural, logical, and correct thing to do.

My experiences inevitably led me to become a palliative medicine physician and an advocate of the philosophy that sometimes enough is enough and death is not always to be feared. While I may admire the persistence and courage of people who are driven to endure any torture to stay alive for a few additional months, weeks, days, or hours, it is difficult for me to fully identify with them. It is also problematic for me to empathize with certain bereaved survivors—family, friends, or fellow medical caregivers—who target and blame an individual or an untoward circumstance for death.

The book represents an honest attempt on my part to listen to and convey all sides of the issues. The topics covered are complex and the bedside decisions resist easy answers. I have striven throughout my writing to be as even-handed as possible and to allow the eloquence and compassion of the different positions to be expressed. However, did I take a position? Absolutely!

So, that is the backdrop as to why I had to write this book—it is all very intimate and very personal. What's more, it would have been a real crime if I had not told the story of these three remarkable nursing staff from my hospital and their conflict over Rosemarie Doherty's death. Amy Gleason, Kim Hoy, and Olga Vasquez represent a profession that goes unheralded. I learned from them just how skewed my vantage point was as a physician, and that it is essential to turn to nurses for a more realistic and multifaceted perspective on end-of-life care. I cannot sufficiently express my gratitude to them for telling me their stories.

Now to the list of people who assisted in this quest. I must start by extolling the incredible forbearance displayed by my wife and sons, who endured my intermittent fits of excitement as well as prolonged compulsion to make sense of this subject. Thanks are also proffered to each member of my extended family—my mother-in-law, Myra Berzoff; my dearest friend, Jaine Darwin; Bob and Cynthia Shilkret; and a sizeable group of nieces and nephews—who patiently listened to my ideas as well as read, improved, and corrected drafts and proofs.

Generous support came from the Rockefeller Foundation, which provided an opportunity to reside at its majestic estate in Bellagio, where I joyfully wrote in a twelfth-century church tower and got to meet a delightful bunch of scholars, poets, artists, and musicians.

The Guggenheim Foundation's fellowship affirmed the importance of the book, opened many doors, and gave me a chance to complete the manuscript in yet another serenely beautiful setting. And I have unbounded gratitude to Tufts University School of Medicine, where I am a professor and which honored me with its Distinguished Faculty Award.

Baystate Health deserves my special thanks. Throughout a quarter century at the Springfield medical center, I have witnessed the organization intelligently grow from a few community hospitals to its current position as the premier medical program in western Massachusetts. Since much of the book's activity takes place at the hospital, the CEO and president, Mark Tolosky, understandably could have worried about public relations and unleashed the lawyers—but instead, the medical center was both encouraging and trusting. Throughout my tenure at Baystate, the administration—especially the senior vice president, Loring Flint (whom I consider to be my Secret Santa)—always graciously supported my research endeavors, while my chairman, Benjamin Liptzin, consistently guided me to fulfill or even exceed my academic aspirations. These individuals were instrumental in granting me an invaluable sabbatical to complete the project, but that sabbatical only became feasible through the extra effort and loyalty of my colleagues, June Plasse, Joyce Smith, Steven Fischel, Adam Mirot, Stephen Luippold, and the other psychiatrists on staff.

Books simply do not get written and published without the counsel and perspicacity of an expert literary agent, and I was extremely fortunate in having Carol Mann's help. I am especially grateful to my skillful and always encouraging original editor, Nancy Miller. Matt Harper then proceeded to marshal the efforts of the many wonderful people at HarperCollins, and he valiantly worked with me to further improve the writing. Matt brought great practical talent, literary proficiency, and diplomatic deftness to this task.

There were three patients whose terminal circumstances and decisions to stop dialysis expressed some of the complicated truths about dying. Each of them entered the public record when their situations resulted in either newspaper articles or the criminal investigation that forms the heart of the book. I honor their memories and am grateful to their families. I hope that any intrusion into their privacy is balanced by the good that can come to others by revealing these events.

In particular, I was pleased to reach Barbara Dilanian's children, Jane and Ken, and to have the opportunity to thank them directly for sharing their private experience. I was also thrilled to connect with several of Benjamin Babcock's friends. Mr. Babcock and his family graciously gave me permission during his last hospitalization to videotape my interviews. He believed that he had some timeless lessons to convey, and I must agree. Rosemarie Doherty's son Jim could not have been more warm and cordial during our calls and e-mails. I greatly appreciate the additional information he and his family offered about her life.

During the prolonged gestation of this book, I tried out a number of titles before the quick-witted and generous writer Suzanne Gordon had a eureka moment and suggested the present one. During the last seven years, I interviewed hundreds of people, and the word *friend* appears with noticeable frequency as an adjective for many of these individuals. In fact, I had to edit out some uses of this word because it became too repetitious. Being a rather private and solitary soul, I was surprised by the realization that I was truly speaking with friends. In this acknowledgment section, I will not repeat the names of those who are directly cited in the text, but I am exceedingly grateful to each of them for their participation.

A considerable number of people were sources of information and supported this project, but their names do not appear in the final text. The following list no doubt reflects some memory lapses on my part and is probably incomplete, but I would like to thank Thomas Szasz, Marilyn August, Judith Nelson, Sharon Weinstein, Dax Cowart, Jay Holtzman, Rorry Zahourek, Nicholas Christakis, Tim Quill, Kathy Foley, Penina and Mickey Glazer, Antonio Artigas Raventós, Luis Cabre, Massimo Antonelli, Ira Byock, Susan Webb, Diane Coleman, Margaret Somerville, Tom Higgins, Gerald Green, Anita Sarro, Frank Marotta, Chad Kollas, Bonnie Steinbock, Michael Collier, Phil and Frannie Levine, Mordechai Halperin, Avraham Steinberg, Avinoam Reches, and Charlie Sabatino. At the same time, I would also like to honor the memories of the late Sharon Smith, who stood up for her Baystate nurses, and Gary Reiter, who introduced many of us to the field of palliative medicine.

I am indebted to Amos Bailey and Deborah Sherman's critiques, and to Y. Michael Barilan's article on Israel's new law, the Patient Nearing Death Act. I also relied on articles and publications by Sharon LaDuke, John Tierney, R. Alonso-Saldivar, John Fauber, Andrew Bard Schmookler, Ann Alpers, Robert Fine, K. Johnson, Cathy Lynn Grossman, K. Severson, Susannah Pugh, Laurie Bobskill, N. Demirsoy, David Tuller, William Kole, D. M. Goldenbaum, Diane Meier, Andrew Billings, Susan Block, L. Cabre, William Yardley, and Laura Landro. Some of the material in Chapter 6, was first published in a medical journal, *Palliative and Supportive Care*, and I appreciate the generosity and good fellowship of its editor, William Breitbart.

There are a few books and a website that I would recommend to anyone interested in learning more about this subject. The American Academy of Hospice and Palliative Medicine has launched www .PalliativeDoctors.org, with information for patients and families. Amos Bailey's *Palliative Response* is being reissued in a new edition and is a practical guide that covers symptoms, treatments, and a number of other pertinent issues. Joan Berzoff and Phyllis Silverman edited *Living with Dying*, a remarkably readable textbook that is written for social workers but understandable to anyone. *Dying Well* by Ira Byock, a former president of the American Academy of Hospice and Palliative Medicine, captures his philosophy that "dying is more than a set of problems to be solved." Robert Orr has just published *Medical Ethics and the Faith Factor*, in which he distills his experience as a physician, ethicist, and man of faith, and manages to lucidly explain how contemporary clinical ethics can be erected on a theological foundation. *The Savage God* by A. Alvarez is still the most interesting book I've encountered on suicide, while Joanne Lynn's *Handbook for Mortals* is the best practical resource for anyone facing the end of life.

The nonfiction author Seth Schulman requires a special acknowledgment. More than any other individual, he provided me with the greatest amount of direct help and positive reinforcement in the writing of this book. Although he scrupulously did his utmost not to inject firm opinions or preferences about the "best" way to write, Seth's hand guided me throughout this process.

The Greenwall Foundation and its president, Bill Stubing, have repeatedly funded my bioethical research investigations. Bill made it possible for me to have the privilege of collaborating in a multicenter study on medical murder accusations with Linda Ganzini, Elizabeth Goy, Jim Cleary, Steve Arons, Nate Goldstein, and Bob Arnold.

There are hundreds of thousands of Americans whose lives are prolonged by dialysis and whose daily accomplishments are rarely acknowledged. I respect how much energy is expended by them and their dialysis staff, and also how much they silently suffer. Nevertheless, I was also completely serious when I wrote of feeling envy for the ability and freedom of dialysis patients to choose when to stop treatment and to have the opportunity to die in a dignified manner. My main professional goal will always be the integration of palliative medicine into the specialty of nephrology, and to thereby discover ways to minimize patient discomfort and maximize the capacity to live life to its fullest.

Finally, I would like to apologize if in publishing this book I unintentionally caused anyone pain or embarrassment. I sincerely tried to understand and accurately convey what I observed and was told during the many interviews, but I may have misconstrued or misinterpreted some of these conversations. If so, please forgive me any discomfort I may have inadvertently caused, and recognize that I did my best to communicate a complicated and intensely personal subject.